"十四五"职业教育国家规划教材

高职高专计算机类专业系列教材——移动应用开发系列

Android 应用开发基础

赖 红 主编

电子工业出版社
Publishing House of Electronics Industry
北京·BEIJING

内 容 简 介

随着移动设备的发展和普及，Android 操作系统越来越得到广泛的应用，目前 Android 操作系统的全球市场份额已接近 90%，在全球范围内占据着主导地位。由于目前 Android 技术发展日新月异，Android 高级开发人才短期将供不应求，Android 高级工程师的就业前景非常紧俏。

本书基于 Google 新推出的 Android Studio 3.5 和 Android SDK10.0 作为开发环境进行编写，全面介绍了 Android 应用开发的相关知识，内容覆盖了 Android 开发环境的搭建、Android 视图、Android 组件、Fragment 组件、广播技术、系统服务、系统内容提供者、多媒体、网络服务和文件数据库管理等。本书以"案例驱动"和"模块化"的方式来讲解 Android 编程的各种理论知识，包含了 31 节理论知识点慕课视频及 50 个实践案例编程操作视频；本书的慕课视频和实践案例视频总时长 720 分钟。这些实践案例能帮助读者更好地理解 Android 各种知识在实际开发中的应用。与本书配套的数字课程已登录"学堂在线"（https://next.xuetangx.com/），读者可以登录网站搜索"Android 应用开发基础"选课进行学习。

本书适合作为高职高专院校计算机软件技术、移动互联应用技术及相关专业的教材，也可作为职业培训的教材或自学者的参考书，还可供从事移动开发的工作者学习参考。

未经许可，不得以任何方式复制或抄袭本书之部分或全部内容。
版权所有，侵权必究。

图书在版编目（CIP）数据

Android 应用开发基础 / 赖红主编. —北京：电子工业出版社，2020.3
ISBN 978-7-121-37592-7

Ⅰ. ①A… Ⅱ. ①赖… Ⅲ. ①移动终端—应用程序—程序设计—高等学校—教材 Ⅳ. ①TN929.53

中国版本图书馆 CIP 数据核字（2019）第 219790 号

责任编辑：贺志洪
印　　刷：三河市鑫金马印装有限公司
装　　订：三河市鑫金马印装有限公司
出版发行：电子工业出版社
　　　　　北京市海淀区万寿路 173 信箱　邮编 100036
开　　本：787×1092　1/16　印张：19.5　字数：499.2 千字
版　　次：2020 年 3 月第 1 版
印　　次：2023 年 9 月第 13 次印刷
定　　价：49.00 元

凡所购买电子工业出版社图书有缺损问题，请向购买书店调换。若书店售缺，请与本社发行部联系，联系及邮购电话：（010）88254888，88258888。
质量投诉请发邮件至 zlts@phei.com.cn，盗版侵权举报请发邮件至 dbqq@phei.com.cn。
本书咨询联系方式：（010）88254609，hzh@phei.com.cn。

源代码和资源说明

本书开发了基于雨课堂的教学 PPT，教师使用雨课堂教学课件可以随时通过微信进行课堂教学；教师通过教材提供的雨课堂教学课件，与学生在课外预习与课堂教学间建立沟通桥梁，让课堂互动永不下线。本书还包含了与课程相关的教学大纲、电子教案、案例源代码、课程练习习题、课程慕课视频和案例实践操作视频。

本书主要包含 50 个案例源代码，源代码按照每章进行组织，本书的源代码结构如下图所示；下面以第 4 章的任务为例来说明一下源代码的命名规则。chapter4：代表第 4 章的源代码目录；chapter4.1.1 代表 4.1 节的第一个案例的源代码。

源代码和资源的访问地址链接：https://pan.baidu.com/s/1ScC2QWWaLoEjhipwBAjGUA；提取码：uey7。

- chapter1.3-logcat
- chapter2.1.1_life
- chapter2.1.2-firstactivity
- chapter2.2-activitymode
- chapter3-middle
- chapter3.1-view
- chapter3.2-button
- chapter3.2-progressbar
- chapter3.2-progressdialog
- chapter3.2-Textview
- chapter3.3-checkbox
- chapter3.3-edittext
- chapter3.3-imageview
- chapter3.3-radiobutton
- chapter3.3-switch
- chapter3.4-baseadapter
- chapter3.4-listview
- chapter3.4-spinner
- chapter4.1.1-fragment
- chapter4.1.2-fragmentdemo
- chapter4.2- viewpagertest
- chapter5-BroadcastBestPractice2
- chapter5.1_PhoneStatus
- chapter5.2_smsreceiver
- chapter5.3_mybroadcast
- chapter5.4-sendselfbroadcast2
- chapter6.2-NotificationTest
- chapter6.3-ServiceTest
- chapter6.4.2-ServiceTest
- chapter6.4.3-ServiceTest
- chapter6.4.4-asynctasktest
- chapter6.4.5-downloadTest
- chapter6_taskmanger
- chapter7.1-RuntimePermissionCall
- chapter7.1-runtimepermissionsms
- chapter7.3-Contact
- chapter7.4-ContactsTest
- chapter7_MedioStoreSDCard
- chapter8-player
- chapter8.1-CameraAlbumTest
- chapter8.2.1-PlayAudioTest
- chapter8.2.2-PlayVideoTest
- chapter9.1.1-WebViewTest
- chapter9.1.2-Httptest
- chapter9.2-okhttptest
- chapter9.3-jsontest
- chapter9.4-volleyhttptest
- chapter10-filesqlite

图　本书的源代码结构

前　　言

　　Android 作为一款由 Google（谷歌）创建的全新的智能手机开发平台，主要应用于移动设备如智能手机、平板电脑、可穿戴设备；从 2008 年 9 月 Google 发布 Android 1.0 到 2019 年 9 月 Google 发布 Android 10.0，Google 几乎以每年一个版本的速度对 Android 进行迭代发布。随着移动设备的发展和普及，Android 操作系统越来越得到广泛的应用，目前 Android 操作系统的全球市场份额已接近 90%，在全球范围内占据着主导地位。由于目前 Android 技术发展日新月异，Android 高级开发人才短期将供不应求，Android 高级工程师的就业前景非常紧俏。Android 系统作为一个成熟的商业移动设备操作系统，本身具有丰富的系统组件、功能强大的四大组件、多媒体、传感器及 SQLite 数据库存储技术，对于很多刚开始学习的初学者来说，由于实践的时间偏少，不容易掌握 Android 开发中的各项技术。

　　本教材紧扣 Android 应用开发岗位知识点需求，对 Android 应用开发工作流程中的知识点进行了提炼和分析，从 Android 基础开始，理论与实践相结合，每个知识点都配备了理论讲解和实践操作案例。通过本教材的学习，可以快速掌握 Android 应用程序开发所需的基础知识，掌握完整的 Android 软件开发的流程和技术架构方法。

　　本教材主要包括：开发环境的搭建、Android 视图、Android 组件、碎片组件、广播技术、系统服务、系统提供者组件、多媒体、网络服务、文件和数据库管理等。

　　本教材包含了 31 节慕课视频及字幕，50 个案例编程慕课视频及字幕源代码；教材案例视频总时长 720 分钟；8 个综合案例和 43 个知识点案例，案例最低兼容 Android 8.0，最高支持 Android 10.0 版本；本教材共包含了 200 多道习题。

　　本教材改进了传统的教学组织模式，将 Android 案例进行功能点的拆分和知识点的分析，通过实例任务进行循序渐进的学习和迭代开发，紧密围绕 Android 程序设计的基础知识和技能，规划设计了 50 个实践任务，涵盖 Android 各个知识点；采用迭代开发的过程，层次递进，循序渐进，每个功能点扩展讲解 Android 的高级应用；每个实践任务都按"任务说明→关键知识及慕课视频→实践任务及操作视频"的结构组织。每个任务以一个生动贴切的实例开头而且实际运行，并给出了详细的一步步的实现步骤和关键知识点的讲解视频，使学生在学习枯燥的 Android 基础理论知识前能够根据提示完成难度适中的程序；在调动学生积极性的同时对程序的结构和流程架构做出分析，引导学生了解实例中涉及的知识点；最后让学生完成一个实例任务，让学生自己练习，熟练地应用刚学习的知识模仿完成，提高学生的成就感和兴趣。

本教材开发了 31 节雨课堂教学课件，教师使用雨课堂教学课件可以随时通过微信进行课堂教学；教师通过教材提供的雨课堂教学课件，和学生在课外预习与课堂教学间建立沟通桥梁，让课堂互动永不下线。使用雨课堂课件，教师可以将带有 MOOC 视频、习题、语音的课前预习课件推送到学生手机，师生沟通可以及时反馈；教师可以通过教学课件让学生在课堂上实时答题、弹幕互动，为传统课堂教学师生互动提供了完美解决方案。教材提供的雨课堂课件科学地覆盖了"课前—课中—课后"的每一个教学环节，为师生提供完整立体的数据支持，个性化报表、自动任务提醒，让教与学更加一目了然。

本教材由赖红担任主编，主要负责全书的组织设计、案例的分析和整体的结构；各章的分工如下，赖红负责第 1~6 章，李钦负责第 7~10 章；联想教育科技有限公司的陈靖女士完成了本书案例的设计和编写工作。

希望本书能帮助 Android 的任课老师将 Android 的开发知识传授给学生，也希望初学者更好地快速掌握 Android 的开发实践技能。由于作者水平有限，疏漏之处在所难免，欢迎广大的读者提出宝贵的意见。作者的联系邮箱：64881623@qq.com。

编者

2020 年 2 月

目 录

1 Android 基础 ··· 1
   1.1 Android 概述 ··· 1
      1.1.1 Android 简介 ·· 1
      1.1.2 实践任务——搭建 Android Studio 开发环境 ·· 5
      1.1.3 单元小测 ··· 12
   1.2 第一个 Android 应用程序 ·· 13
      1.2.1 编写第一个 Android 应用程序 ·· 13
      1.2.2 Android 应用程序结构 ··· 18
      1.2.3 单元小测 ··· 21
   1.3 Android 应用程序调试 ·· 22
      1.3.1 Android 调试工具 ·· 22
      1.3.2 Android 调试实现 ·· 24
      1.3.3 单元小测 ··· 27

2 Android 视图 ·· 29
   2.1 Android 视图概述 ··· 29
      2.1.1 Activity ·· 29
      2.1.2 Activity 生命周期实例 ··· 35
      2.1.3 Activity 数据传递实例 ··· 38
      2.1.4 单元小测 ··· 47
   2.2 Android 启动模式 ··· 48
      2.2.1 standard ··· 49
      2.2.2 singleTop ·· 49
      2.2.3 singleTask ··· 50
      2.2.4 singleInstance ··· 51
      2.2.5 单元小测 ··· 52

3 Android 布局与组件 ·· 55
   3.1 Android 布局 ·· 55

- 3.1.1 绝对布局 ··· 58
- 3.1.2 相对布局 ··· 58
- 3.1.3 线性布局 ··· 62
- 3.1.4 约束布局 ··· 66
- 3.1.5 单元小测 ··· 75
- 3.2 Android 基础组件 ··· 76
  - 3.2.1 TextView ··· 76
  - 3.2.2 Button ··· 79
  - 3.2.3 EditText ··· 81
  - 3.2.4 ProgressBar ··· 83
  - 3.2.5 单线程模型 ··· 85
  - 3.2.6 单元小测 ··· 88
- 3.3 Android 中级组件 ··· 89
  - 3.3.1 CheckBox ··· 89
  - 3.3.2 Switch ··· 91
  - 3.3.3 RadioButton ··· 93
  - 3.3.4 ImageView ··· 96
  - 3.3.5 单元小测 ··· 101
- 3.4 Android 适配器 ··· 103
  - 3.4.1 Adapter 适配器 ··· 103
  - 3.4.2 Spinner ··· 104
  - 3.4.3 ListView ··· 106
  - 3.4.4 自定义 Adapter ··· 110
  - 3.4.5 单元小测 ··· 115

4 Android Fragment ··· 117

- 4.1 Fragment 组件 ··· 117
  - 4.1.1 Fragment 的生命周期 ··· 118
  - 4.1.2 Fragment 通信 ··· 127
  - 4.1.3 Fragment 动态加载 ··· 131
  - 4.1.4 单元小测 ··· 136
- 4.2 ViewPager 组件 ··· 138
  - 4.2.1 ViewPager 概述 ··· 138
  - 4.2.2 引导页与选项卡 ··· 143
  - 4.2.3 单元小测 ··· 149

5 Android 广播 ··· 151

- 5.1 广播概述 ··· 151
  - 5.1.1 Android 广播收发机制 ··· 151
  - 5.1.2 实践案例——获取设备中电池的电量 ··· 154

  5.1.3 单元小测 ·· 159
5.2 广播收发机制 ·· 160
  5.2.1 知识点讲解——广播收发机制 ····················· 160
  5.2.2 实践案例——显示网格状态 ························ 162
  5.2.3 单元小测 ·· 165
5.3 自定义广播 ·· 166
  5.3.1 知识点讲解——自定义广播 ························ 166
  5.3.2 实践案例——实现自定义广播 ····················· 167
  5.3.3 单元小测 ·· 169
5.4 有序广播 ·· 169
  5.4.1 知识点讲解——有序广播 ···························· 169
  5.4.2 实践案例——实现有序广播 ························ 171
  5.4.3 单元小测 ·· 172

6 Android 系统服务 ··· 174

6.1 系统服务概述 ·· 174
  6.1.1 Android 的服务组件 ···································· 174
  6.1.2 单元小测 ·· 177
6.2 访问系统服务 ·· 178
  6.2.1 Android 的系统服务组件 ····························· 178
  6.2.2 实践案例——实现通知服务 ························ 179
  6.2.3 单元小测 ·· 180
6.3 自定义服务 ·· 181
  6.3.1 知识点讲解——自定义服务 ························ 181
  6.3.2 实践案例——调用 Service 生命周期函数 ···· 182
  6.3.3 单元小测 ·· 186
6.4 多线程 ··· 187
  6.4.1 知识点讲解——多线程 ······························· 187
  6.4.2 实践案例——创建多线程服务 ····················· 188
  6.4.3 单元小测 ·· 192

7 Android 内容提供者 ·· 194

7.1 Android 运行权限 ·· 194
  7.1.1 知识点讲解——运行权限 ···························· 194
  7.1.2 实践案例——设置电话权限 ························ 196
  7.1.3 单元小测 ·· 198
7.2 URL 和 URI 概述 ·· 198
  7.2.1 URL 和 URI ··· 198
  7.2.2 单元小测 ·· 199
7.3 ContentProvider ·· 200

7.3.1　知识点讲解——ContentProvider ································· 200
　　　7.3.2　实践案例——读取联系人 ·· 202
　　　7.3.3　单元小测 ·· 205
　7.4　访问通讯录 ·· 206
　　　7.4.1　知识点讲解——访问通讯录 ····································· 206
　　　7.4.2　实践案例——访问通讯录 ····································· 207
　　　7.4.3　单元小测 ·· 212

8　多媒体 ··· 214

　8.1　拍照服务 ·· 214
　　　8.1.1　知识点讲解——拍照服务 ·· 214
　　　8.1.2　实践案例——拍照服务 ·· 215
　　　8.1.3　单元小测 ·· 219
　8.2　音视频服务 ·· 220
　　　8.2.1　知识点讲解——音视频服务 ····································· 220
　　　8.2.2　实践案例——视频播放 ·· 221
　　　8.2.3　单元小测 ·· 227

9　网络服务 ·· 230

　9.1　网络服务概述 ·· 230
　　　9.1.1　知识点讲解——网络服务 ·· 230
　　　9.1.2　实践案例——使用 HTTP 协议访问网络 ······················ 233
　　　9.1.3　单元小测 ·· 239
　9.2　网络框架 ·· 241
　　　9.2.1　知识点讲解——网络框架 ·· 241
　　　9.2.2　实践案例——网络框架 ·· 242
　　　9.2.3　单元小测 ·· 245
　9.3　JSON 协议 ·· 246
　　　9.3.1　知识点讲解——JSON 协议 ······································ 246
　　　9.3.2　实践案例——访问天气实例的应用 ···························· 250
　　　9.3.3　单元小测 ·· 256
　9.4　Volley ·· 257
　　　9.4.1　知识点讲解——Volley 网络框架 ······························· 257
　　　9.4.2　实践案例——使用 Volley 框架实现天气预报的应用 ······ 257
　　　9.4.3　单元小测 ·· 262

10　数据存储 ·· 264

　10.1　文件存储 ·· 264
　　　10.1.1　文件保存 ·· 265
　　　10.1.2　文件读取 ·· 272
　　　10.1.3　SharePreferences 存储 ·· 277

	10.1.4 单元小测 ···	284
10.2	数据库存储 ···	286
	10.2.1 知识点讲解——嵌入式数据库 SQLite ··	286
	10.2.2 实践案例——将个人信息存储到 SQLite 数据库 ··	287
	10.2.3 单元小测 ···	297

参考文献 ·· 300

1 Android 基础

 知识点

Android 的历史和架构、Android 的集成开发环境 Android Studio、Android 程序调试。

 能力点

1. 熟练掌握 Android 架构。
2. 搭建 Android 的集成开发环境 Android Studio。
3. 使用 Android Studio 开发第一个应用程序。

 ## 1.1 Android 概述

1.1.1 Android 简介

Android 概述
（慕课）

1. Android 版本

Android 一词的本义指"机器人"，同时也是 Google 于 2007 年 11 月 5 日宣布的基于 Linux 平台的开源手机操作系统的名称，该平台由操作系统、中间件、用户界面和应用软件组成。

2003 年 10 月，Andy Rubin 等人创建 Android 公司，并组建 Android 团队。2005 年 8 月 17 日，Google 低调收购了成立仅 22 个月的高科技企业 Android 及其团队。Andy Rubin 成为 Google 公司工程部副总裁，继续负责 Android 项目。2007 年 11 月 5 日，谷歌公司正式向外界展示了这款名为 Android 的操作系统，并且在这天谷歌宣布建立一个全球性的联盟组织，该组织由 34 家手机制造商、软件开发商、电信运营商及芯片制造商共同组成，并与 84 家硬件制造商、软件开发商及电信营运商组成开放手持设备联盟（Open Handset Alliance）来共同研发和改良 Android 系统，这一联盟将支持谷歌发布的手机操作系统及应用软件，Google 以 Apache 免费开源许可证的授权方式，发布了 Android 的源代码。2008 年，在 Google I/O 大会上，谷歌提出了 Android

HAL 架构图，在同年的 8 月 18 日，Android 获得了美国联邦通信委员会（FCC）的批准，在 2008 年 9 月，谷歌正式发布了 Android 1.0 系统，这也是 Android 系统最早的版本。

  2009 年 4 月，谷歌正式推出了 Android 1.5 系统，从 Android 1.5 版本开始，谷歌开始将 Android 的版本以甜品的名字命名，Android 1.5 命名为 Cupcake（纸杯蛋糕）。该系统与 Android 1.0 相比有了很大的改进。2009 年 9 月，谷歌发布了 Android 1.6 正式版，并且推出了搭载 Android 1.6 正式版的手机 HTC Hero（G3），凭借着出色的外观设计及全新的 Android 1.6 操作系统，HTC Hero（G3）成为当时全球最受欢迎的手机。Android 1.6 也有一个有趣的甜品名称，它被称为 Donut（甜甜圈）。2010 年 2 月，Linux 内核开发者 Greg Kroah-Hartman 将 Android 的驱动程序从 Linux 内核"状态树"（"staging tree"）上除去，从此，Android 与 Linux 开发主流分道扬镳。在同年的 5 月，谷歌正式发布了 Android 2.2 操作系统。谷歌将 Android 2.2 操作系统命名为 Froyo，翻译为冻酸奶。2010 年 10 月，谷歌宣布 Android 系统达到了第一个里程碑，即电子市场上获得官方数字认证的 Android 应用数量已经达到了 10 万个，Android 系统的应用增长非常迅速。2010 年 12 月，谷歌正式发布了 Android 2.3 操作系统 Gingerbread（姜饼）。2011 年 1 月，谷歌称每日的 Android 设备新用户数量达到了 30 万部，到 2011 年 7 月，这个数字增长到 55 万部，而 Android 系统设备的用户总数达到了 1.35 亿部，Android 系统已经成为智能手机领域占有量最高的系统。2011 年 8 月 2 日，Android 手机已占据全球智能机市场 48% 的份额，并在亚太地区市场占据统治地位，终结了 Symbian（塞班系统）的霸主地位，跃居全球第一。2011 年 9 月，Android 系统的应用数目已经达到了 48 万，而在智能手机市场，Android 系统的占有率已经达到了 43%，继续排在移动操作系统的首位，随后谷歌发布了全新的 Android 4.0 操作系统，这款系统被谷歌命名为 Ice Cream Sandwich（冰激凌三明治）。

  2012 年 1 月 6 日，谷歌 Android Market 已有 10 万开发者推出超过 40 万个活跃的应用程序，大多数的应用程序为免费程序。Android Market 应用程序商店目录在新年首周周末突破 40 万基准，距离突破 30 万应用仅 4 个月。在 2011 年的早些时候，Android Market 从 20 万增加到 30 万应用也花了 4 个月。2012 年 10 月 30 日，Google 发布 Android 4.2 Jelly Bean 原生系统用户界面，Android 4.2 沿用"果冻豆"这一名称，以反映这种最新操作系统与 Android 4.1 的相似性，但 Android 4.2 推出了一些重大的新特性，具体包括 Photo Sphere 全景拍照功能；键盘手势输入功能；改进锁屏功能，包括锁屏状态下支持桌面挂件和直接打开照相功能等；可扩展通知，允许用户直接打开应用；Gmail 邮件可缩放显示；Daydream 屏幕保护程序；用户连点三次可放大整个显示屏，还可用两根手指进行旋转和缩放显示，以及专为盲人用户设计的语音输出和手势模式导航功能等；支持 Miracast 无线显示共享功能；Google Now 允许用户使用 Gmail 作为新的数据来源，如改进后的航班追踪功能、酒店和餐厅预订功能及音乐和电影推荐功能等。

  2014 年 6 月 25 日，Google I/O 大会上发布了 Developer 版（Android 5.0 Lollipop），之后在 2014 年 10 月 15 日正式发布 Lollipop（棒棒糖）。

  2016 年 5 月 18 日，Google I/O 大会上发布了 Android 7.0，支持 Nexus 6P、Nexus 5X、Nexus 6、Pixel C、Nexus 9 及 Nexus Player 这几款设备。Android 7.0 新功能以实用为主，比如分屏多任务、全新设计的通知控制栏等。

  2017 年 8 月 22 日，Google I/O 大会上发布了 Android 8.0 Oreo（奥利奥），加大了对 App 在后台操作的限制。这种限制在一定程度上延长了安卓机在"睡眠"（Doze）模式下电池的续航能力，它让不在使用的 App 进入睡眠状态，使用时再唤醒。它要达到的目标是在不卸载程序、不改变用户使用习惯的情况下，减少后台应用的耗电。同时，这种对后台应用的限制也会加快运行的速度。

2018年8月7日上午，谷歌正式发布Android 9.0正式版系统，并宣布系统版本Android P的正式命名为代号"Pie"。Android 9.0新增了支持类似于iPhone X的刘海屏设计，具体体现为优化屏幕内容显示，能够让系统或者应用充分利用整块屏幕，尤其是两只"猫耳朵"位置。在Android 9.0系统当中，谷歌还会进一步将谷歌助手集成到应用中、进一步优化电池续航、支持多屏和可折叠屏等。

2019年9月3日，谷歌发布了Android 10正式版。Android 10推出了全新的更新模式，不需要重启设备就能实时进行系统更新，这对于手机系统来说史无前例。Android 10基于语音识别技术，不需要联网就可以将视频中的语音实时转化成字幕，同时新增了手势导航、屏幕左滑后退的功能。另外Android 10系统增加了对折叠屏的支持，如分屏操作，应用可以在手机屏幕打开和折叠时无缝切换，同时系统还增加了对5G网络的支持。

截至2019年第三季度，安卓系统占据了智能手机操作系统市场的85.9%，其中中国市场的占有率接近90%；谷歌旗下应用商店Google Play应用数量已达360万款，开发者达35万人。过去五年中谷歌共向开发者支付了90亿美元，谷歌应用商店下载量累计达700亿次。

2. Android的体系结构

安卓作为一个优越稳定的平台背后必有一个成熟的系统架构，Android系统架构从底向上一共分了4层，从上层到下层分别是应用程序层、应用程序框架层、系统运行库层及Linux内核层；每一层都把底层实现封装，并暴露调用接口给上一层。Android的体系结构如图1-1所示。

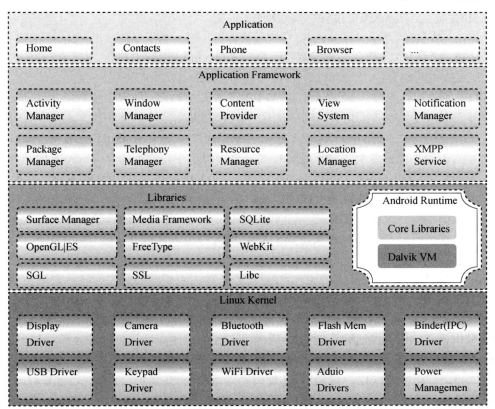

图1-1 Android的体系结构

（1）Linux 内核层（Linux Kernel）

Android 是基于 Linux2.6 内核的，其核心系统服务如安全性、内存管理、进程管理、网络协议及驱动模型都依赖于 Linux 内核。Android 运行在 Linux Kernel 2.6 之上，但是 Linux 内核受 GNU 协议约束的部分已经进行了替换，在 Android 基础上的程序可以用于商业目的。Linux 内核是硬件和软件层之间的抽象层。

（2）系统运行库层

系统运行库层包括两部分：系统核心库和安卓运行时（Libraries & Android Runtime）。

系统核心库是应用程序框架的支撑，是连接应用程序框架层与 Linux 内核层的重要纽带。其主要包括如下组件。

- Surface Manager：显示系统管理库，负责把 2D 或 3D 内容显示到屏幕；在执行多个应用程序时，负责管理显示与存取操作间的互动。
- Media Framework：多媒体库，负责支持图像，支持多种视频和音频的录制与回放；基于 PacketVideo OpenCore；支持多种常用的音频、视频格式录制和回放，编码格式包括 MPEG4、MP3、H.264、AAC、ARM。
- SQLite：一个功能强大的轻量级嵌入式关系数据库。
- OpenGL|ES：根据 OpenGL ES 1.0API 标准实现的 3D 绘图函数库。
- FreeType：提供点阵字和向量字的描绘与显示。
- WebKit：一套网页浏览器的软件引擎。
- SGL：底层的 2D 图形渲染引擎。
- SSL：在 Android 通信过程中实现握手。
- Libc：从 BSD 继承来的标准 C 系统函数库，专门为基于 embedded Linux 的设备定制。

Android 应用程序采用 Java 语言编写，程序在 Android 运行中执行，其运行时分为核心库和 Dalvik 虚拟机两部分。

- 核心库：核心库提供了 Java 语言 API 中的大多数功能，同时也包含了 Android 的一些核心 API，如 android.os、android.net、android.media 等。
- Dalvik 虚拟机：Android 程序不同于 J2ME 程序，每个 Android 应用程序都有一个专有的进程，并且多个程序不是运行在一个虚拟机中，而是每个 Android 程序都有一个 Dalvik 虚拟机的实例，并在该实例中执行。Dalvik 虚拟机是一种基于寄存器的 Java 虚拟机，而不是传统的基于栈的虚拟机，并进行了内存资源使用的优化及支持多个虚拟机的特点。需要注意的是，不同于 J2ME，Android 程序在虚拟机中执行的并非编译后的字节码，而是通过转换工具 DX 将 Java 字节码转成 dex 格式的中间码。

（3）应用程序框架层（Application Framework）

应用程序框架层是我们从事 Android 开发的基础，很多核心应用程序也是通过这一层来实现其核心功能的，该层简化了组件的重用，开发人员可以直接使用其提供的组件来进行快速的应用程序开发，也可以通过继承而实现个性化的拓展。其主要包括如下组件。

- Activity Manager（活动管理器）：管理各个应用程序生命周期及通常的导航回退功能。
- Window Manager（窗口管理器）：管理所有的窗口程序。
- Content Provider（内容提供器）：使得不同应用程序之间存取或者分享数据。
- View System（视图系统）：构建应用程序的基本组件。
- Notification Manager（通告管理器）：使得应用程序可以在状态栏中显示自定义的提示

信息。
- Package Manager（包管理器）：Android 系统内的程序管理。
- Telephony Manager（电话管理器）：管理所有的移动设备功能。
- Resource Manager（资源管理器）：提供应用程序使用的各种非代码资源，如本地化字符串、图片、布局文件、颜色文件等。
- Location Manager（位置管理器）：提供位置服务。
- XMPP Service（XMPP 服务）：提供 Google Talk 服务。

（4）应用程序层（Application）

Android 平台不仅仅是操作系统，也包含了许多应用程序，诸如 SMS 短信客户端程序、电话拨号程序、图片浏览器、Web 浏览器等。这些应用程序都是用 Java 语言编写的，并且这些应用程序都是可以被开发人员开发的其他应用程序所替换，这点不同于其他手机操作系统固化在系统内部的系统软件，更加灵活和个性化。

1.1.2　实践任务——搭建 Android Studio 开发环境

1. JDK 的下载、安装和环境配置

JDK 的全称是 Java（TM）SE Development Kit，即 Java 标准版（Standard Edition）开发工具包。这是 Java 开发和运行的基本平台。换句话说所有用 Java 语言编写的程序要运行都离不开它，而用它就可以编译 Java 代码为类文件。注意，不要下载 JRE（Java Runtime Environment，Java 运行时环境），因为 JRE 不包含 Java 编译器和 JDK 类的源码。

Android Studio
安装与配置
（实践案例）

（1）JDK 下载

进入 Oracle 公司的主界面（http://www.oracle.com/index.html），选择"Downloads"页面中的"Java for Developer"，如图 1-2 所示。

图 1-2　Oracle 公司的主界面"Downloads"页面中的"Java for Developer"下载

进入下载页面后，单击"JDK DOWNLOAD"按钮，在 JDK 下载页面中单击"Accept License Agreement"按钮，进入 JDK 的下载列表（见图 1-3），根据操作系统的不同选择不同的 JDK 版本（32 位操作系统选择 Windows x86；64 位操作系统选择 Windows x64）。

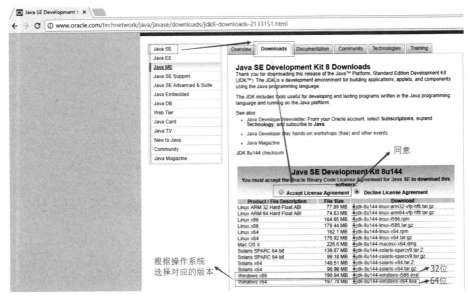

图 1-3　根据操作系统选择 JDK 下载版本

（2）JDK 安装

单击下载完成的 JDK 可执行文件 jdk-8u144-windows-x86.exe（本书下载的是 64 位 Windows x86 版本），单击"下一步"按钮（见图 1-4），选择安装的 JDK 路径（见图 1-5），完成安装。

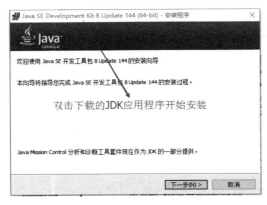

图 1-4　JDK 安装

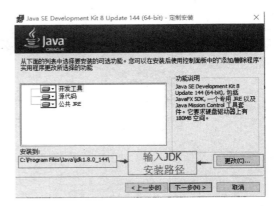

图 1-5　JDK 安装路径设置

（3）JDK 的环境变量配置

在桌面上用鼠标右键单击"我的电脑"图标，在弹出的快捷菜单中选择"属性"命令，进入计算机属性设置页面；在计算机属性设置页面中，单击左边侧栏中的"高级系统设置"项（见图 1-6），在打开的"系统属性"对话框的"高级"选项卡中，单击"环境变量"按钮，如图 1-7 所示。

图 1-6　计算机属性设置页面　　　　　图 1-7　"系统属性"对话框

在"环境变量"设置页，新建 JAVA_HOME 和 Path 两个用户环境变量。方法是：在"Administrator 的用户变量"栏中单击"新建"按钮，在打开的"新建用户变量"窗口的"变量名"中输入"JAVA_HOME"，"变量值"设置为 JDK 的安装路径（本书的安装路径为 C:\Program Files\Java\jdk1.8.0_144），如图 1-8 所示。按照同样的方法，新建 Path 的用户环境变量："变量名"为"Path"，变量值设置为 JDK 的安装路径下面的 bin 子目录（本书的安装路径为 C:\Program Files\Java\ jdk1.8.0_144\bin），如图 1-9 所示。

图 1-8　设置 JAVA_HOME 用户环境变量　　　　图 1-9　设置 Path 用户环境变量

（4）验证是否成功

按快捷键 WIN+R，在命令行中输入"CMD"命令（见图 1-10），进入命令行显示界面。

图 1-10　WIN 键+R 键输入 CMD 命令

在命令行界面中分别输入 javac 和 java 命令，若出现的画面分别如图 1-11 和图 1-12 所示，则软件安装配置成功；若不同，请检查环境变量里变量值的设置是否正确。

图 1-11　CMD 命令行中输入 javac 命令　　　　图 1-12　CMD 命令行中输入 java 命令

2．Android Studio 开发环境的下载和安装

（1）Android Studio 开发环境的下载

谷歌提供了安卓集成的 Android Studio 开发环境，可以从官网和中文社区中下载，本书的开发环境版本是 2019 年 10 月发布的 Android Studio 3.5.1。

登录安卓的开发网站（https://developer.android.google.cn/studio），选择"DOWNLOAD ANDROID STUDIO"下载软件，如图 1-13 所示。

图 1-13　Android Studio 开发环境下载

（2）Android Studio 开发环境的安装

下载成功后打开软件包，单击运行 android-studio.exe 就可以安装开发环境，如图 1-14 所示。

图 1-14　Android Studio 开发环境的安装

首先进入 Android Studio Setup 界面，如图 1-15 所示，单击"Next"按钮进入选择组件页面，如图 1-16 所示；在选择组件页面中，将所需要安装的组件全部选中，单击"Next"按钮进入同意协议页面。

图 1-15　Android Studio Setup 界面

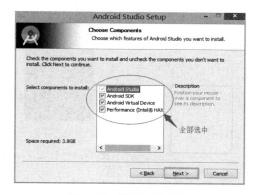

图 1-16　选择组件页面

在同意协议页面中单击"I Agree"（同意）按钮，如图 1-17 所示。进入目录设置页面，设置 Android Studio 的安装路径和 Android SDK 的安装路径，如图 1-18 所示。

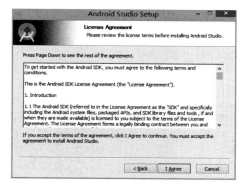

图 1-17　同意协议页面

图 1-18　目录设置页面

设置完成 Android Studio 的安装路径和 Android SDK 的安装路径后，单击"Next"按钮开始安装，如图 1-19 所示，大概需要 3 分钟左右，安装完成，如图 1-20 所示。

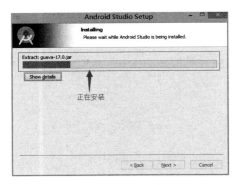
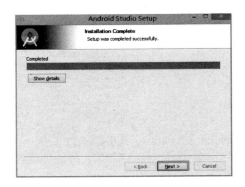

图 1-19　开始安装页面　　　　　　　　　图 1-20　安装过程页面

安装完成后出现启动 Android Studio 的复选框，如图 1-21 所示，安装顺利完成。

图 1-21　完成 Android Studio 开发环境的安装

（3）Android Studio 开发环境

单击启动 Android Studio 开发环境，进入开发环境界面，如图 1-22 所示。

Android Studio 开发环境主要包括工作空间、项目视图、透视图等，整体的布局如图 1-22 所示。

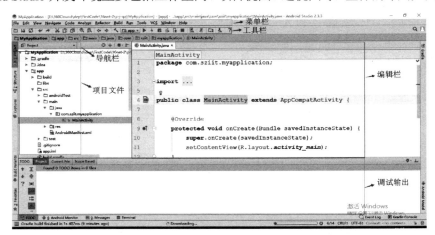

图 1-22　Android Studio 开发环境启动界面

工作空间（Workspace）：工作空间绑定了一个目录作为默认的工作路径，所有在该工作空间新建的项目和文件，默认均保存在该目录下，所有用户的定制选择也保持在这个目录下。

项目（Project）：项目是我们进行开发的工作对象，在一个工作空间中，可以创建多个项目。

透视图（Perspective）：透视图定义了工作台上的一组视图（Views）、编辑器的初始集合和布局显示。针对用于完成某个特定类型的任务或工作，透视图提供一组功能集合，如 Java 透视图、DDMS 调试视图。一般透视图都是由一个编辑器和一个或多个视图组成的，如本书中的 Java 透视图由导航视图、编辑视图和输出调试视图组成。

3. AVD 模拟器的设置与使用

AVD 全称 Android Virtural Device（Android 模拟器），在 Android SDK 1.5 版本后的 Android 开发中，如果需要使用虚拟移动设备必须至少创建 1 个 AVD。打开基于 Android Studio 的开发环境，按以下步骤创建 AVD。

（1）在菜单栏中选择"Tools（工具）"→"Android AVD Manager"命令进入安卓虚拟设备管理界面，如图 1-23 所示。单击"Create Virtual Device"（创建新设备）按钮，在打开的界面的硬件选择栏中，选择手机设备，并为手机设备选择一个型号和屏幕尺寸，单击"Next"按钮，如图 1-24 所示。

图 1-23　安卓虚拟设备管理界面　　　　　　图 1-24　新建 AVD 手机设备

在"系统镜像"栏中，为新建的安卓虚拟设备选择合适的版本，如图 1-25 所示；如果版本已下载，可以直接选择；如果版本没有下载，需要在线下载 Android 版本；版本选择完成后单击"Next"按钮进入模拟器列表界面；可以查看已建好的模拟器列表，如图 1-26 所示。

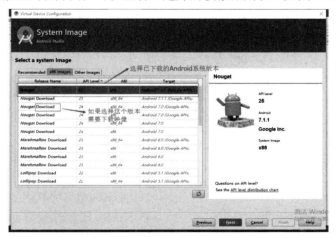

图 1-25　AVD 系统版本管理

（2）创建安卓虚拟设备后，在模拟器列表中运行模拟器；运行创建的"Nexus 5"虚拟设备，如图 1-27 所示。

图 1-26　模拟器列表界面

图 1-27　Android AVD Manager 启动虚拟设备

1.1.3　单元小测

单选题：

1. Android1.0 是 Google 在（　　）年 9 月发布的。
A. 2009　　　　　　B. 2010　　　　　　C. 2008　　　　　　D. 2011
2. Android Oreo 对应的版本是（　　）。
A. 5.0　　　　　　　B. 6.0　　　　　　　C. 7.0　　　　　　　D. 8.0
3. Android 是 Google 公司基于（　　）平台开发的手机的操作系统。
A. Linux　　　　　　B. Windows　　　　C. Mac　　　　　　D. UNIX
4. Android 的分层架构中，应用程序层采用（　　）语言进行开发。
A. C　　　　　　　　B. Python　　　　　C. Java　　　　　　D. C++
5. Android 开发应用程序主要采用（　　）层提供的接口进行开发。
A. Linux 内核层　　　B. 系统运行库层　　C. 应用程序框架层　D. 应用程序层
6. Java 开发工具包（Java Development kit，JDK）是由哪个公司开发的？（　　）
A. Google（谷歌）　　　　　　　　　　　B. MicroSoft（微软）
C. Oracle（甲骨文）　　　　　　　　　　D. IBM（国际商业机器公司）
7. AS 开发工具包（Android Studio）是由哪个公司开发的？（　　）
A. Google（谷歌）　　　　　　　　　　　B. MicroSoft（微软）
C. Oracle（甲骨文）　　　　　　　　　　D. IBM（国际商业机器公司）

1.2　第一个 Android 应用程序

1.2.1　编写第一个 Android 应用程序

Android 程序
（慕课）

本小节的任务是编写一个 Android 的应用程序，程序启动后在 Android 手机上或者模拟设备上显示"Welcome to Android"的信息，完成本任务后，程序的运行结果如图 1-28 所示。

第一个 Android
应用程序
（实践案例）

图 1-28　"Welcome to Android"应用程序显示图

（1）在开发工具环境视图"包资源管理器"中右击，在弹出的快捷菜单中选择"新建"→"Android Application Project"命令，在 Android Studio 开发环境的欢迎界面中单击"Start a new Android Studio Project"，进入创建新项目的界面，如图 1-29 所示。

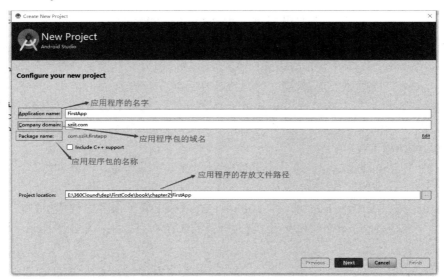

图 1-29　创建新项目的界面

Application name：用户看到的应用名称，在这个项目中，名称是"FirstApp"。

Company domain：表示公司域名，如果是个人开发者，或者没有公司域名，可以像本书一样填写 sziit.com 就可以了。

Package name：应用程序的包名（在 Java 程序设计语言中遵循同样的包名）。这个包名不能与 Android 系统中的其他应用的包名重复，包名具有唯一性。因此，包名通常使用你的单位或组织的反向域名命名。以这个项目为例，Android Studio 会根据应用的名称和公司域名来自动生成合适的包名，比如本例子的包名为 com.sziit.firstapp。

Projection location：项目存放的文件夹路径，读者可以根据自己的情况选择。

（2）单击"Next"按钮，在打开的界面中对项目的目标设备进行设置，如图 1-30 所示。

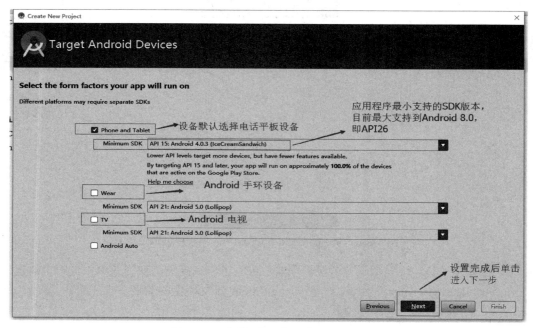

图 1-30　对项目的目标设备进行设置

Minimum SDK：应用能支持的最低 Android 平台版本，表示正在使用的 API level。为了支持更多的设备，应当是支持的 Android 平台版本越低越好，从而让你的应用程序能够提供低版本核心功能集。如果你的应用程序的一些特性只能在新的 Android 平台上面支持，那么你可以让这些特性只有在支持的 Android 平台版本上面运行的时候体现（更多了解支持不同平台版本）。由于 Android 4.0 以上的设备占据了将近 99%的市场份额，在本项目中这一项选择默认的设定值 API 15。

Wear、TV 和 Android Auto，这几个选项分别代表可穿戴设备、电视和汽车，目前国内市场应用的数量并不是很多；设置完成后单击"Next"按钮进入视图设置界面。

（3）视图设置界面如图 1-31 所示，Android Studio 提供了很多内置模板，在最开始的学习阶段，我们可以选择 Empty Activity 创建一个新的视图就可以了；单击"Next"按钮进入下一个界面。

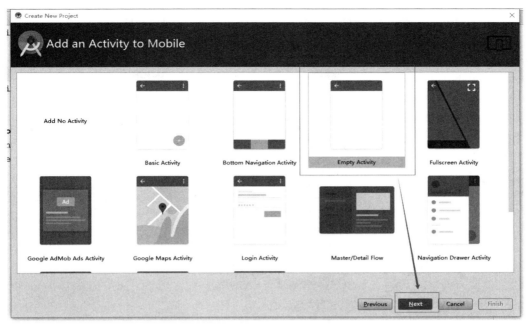

图 1-31　视图设置界面

（4）在视图和布局命名界面中，可以给新建的视图和布局进行命名，如图 1-32 所示。其中"Activity Name"代表活动的名字，使用默认的"MainActivity"；"Layout Name"代表布局的名字，使用默认的"activity_main"；单击"Finish"按钮，项目创建成功。

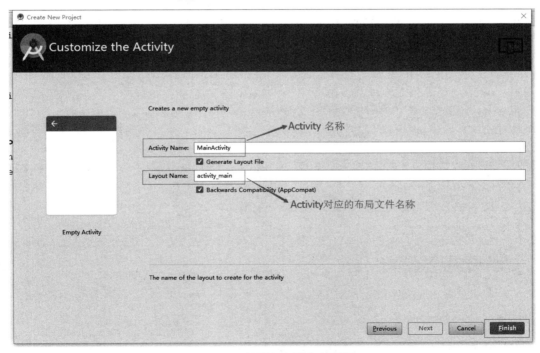

图 1-32　视图和布局命名界面

（5）上一节中已经新建完成了 FirstApp 的项目，需要对项目中的一些内容进行设置，比如改变显示"HelloWorld"的内容为"Welcome to Android"。操作步骤如下：

在开发工具环境视图"包资源管理器"中选择"res\values\strings.xml"文件，再选择"Open Editor"模式，新建"act_main_welcome"变量，修改变量的"Default Value"为"Welcome to Android"，如图 1-33 所示。

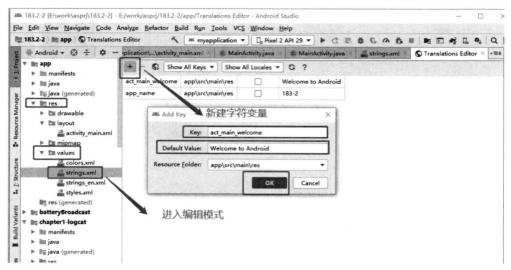

图 1-33　新建"act_main_welcome"变量

选择"res\layout\activity_main.xml"文件，选择文件的"Design"模式，单击"TextView"控件，在"Properties"中单击"Text"项，在弹出的"Pick a Resource"（选择资源）对话框中，选择"Text"项对应的变量为"act_main_welcome"变量，如图 1-34 所示。

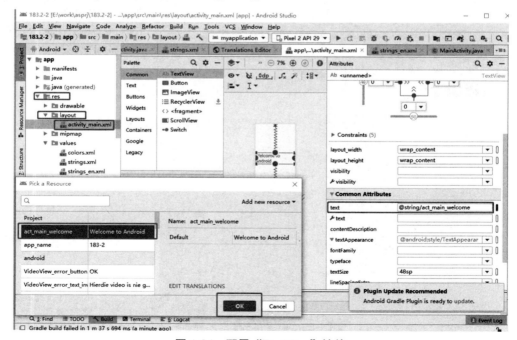

图 1-34　配置"TextView"控件

（6）在开发工具环境视图的工具栏中，选中"app"Module，单击"运行"按钮，进入运行配置设置界面，如图1-35所示。

在运行配置 Target（目标）标签页，编写的程序可以运行在所有的实际设备和虚拟设备中，每次运行的时候用户可以自由选择。系统选择已运行的实际和虚拟设备后单击"OK"按钮，编写的应用程序可以运行到模拟器或者手机设备上。

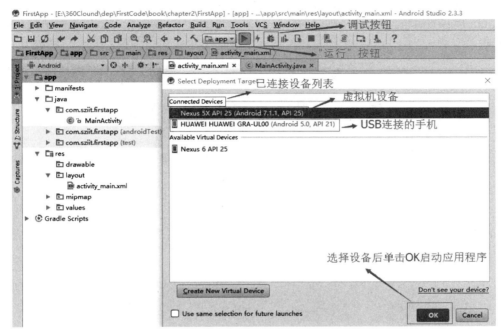

图 1-35　运行配置设置界面

如果选择"Available Virtual Devices"单选按钮中的 Android 模拟器"Nexus 6 API 25"，Android Studio 会重新启动一个模拟器，界面显示出第一个 Android 运行程序"Welcome to Android"，如图 1-36 所示。

图 1-36　Android 模拟器运行程序

1.2.2 Android 应用程序结构

1. Android 应用程序文件结构

在开发工具环境视图"包资源管理器"中单击创建的项目"app",项目中包含的文件结构如图 1-37 所示。Android Studio 主要支持"Project"、"Android"和"Packages"三种方式;其中 Android 结构最简洁,每个项目可以支持多个 Module,每一个 Module 都是独立的一个应用程序。

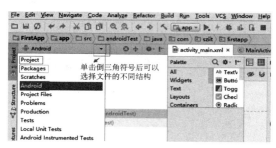

图 1-37 项目中包含的文件结构

Android Studio 主要使用"Android"方式管理项目文件,"Android"项目的文件结构如图 1-38 所示。

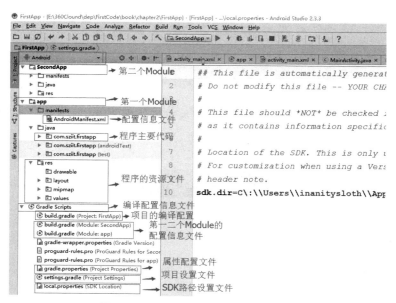

图 1-38 "Android"项目的文件结构

项目主要包含的文件夹主要有以下几个。
- ◆ AndroidManifest.xml:该文件是功能清单文件,列出了应用中所使用的所有组件,如"activity",以及后面要学习的广播接收者、服务等组件。
- ◆ java/:专门存放编写的 Java 源代码的包。
- ◆ res/:res 是 resource 的缩写,该目录为资源目录,可以存放一些图标、界面文件、应用中用到的文字信息。
- ◆ build.gradle：项目的编译配置文件保存在 Project 文件夹的 build.gradle 文件中；Module 的编译配置文件保存在 Module 文件夹的 build.gradle 文件属性配置文件中；项目的属性配置文件保存在 Project 文件夹的 gradle.properties 文件中；项目的设置文件保存在 Project 文件夹的 setting.gradle 文件中；项目的 SDK 路径，设置文件保存在 Project 文件夹的 local.properties 文件中。

Android Studio 使用 Gradle 工具进行程序的构建和编译；Gradle 构建文件包括了 Project\setting.gradle、Project\build.gradle 和 Module\build.gradle。Android 项目创建成功后会自动下载和更新 Gradle。下面是项目的编译配置文件，如图 1-39 所示。

```
buildscript {                //设置项目的编译环境
        repositories {       //支持 java 依赖库管理,用于项目的依赖
              jcenter()
        }
        dependencies {
              classpath 'com.android.tools.build:gradle:3.1.4'    //Grade 工具版本
} //多项目的集中配置，对于 Module 的配置，都是基于项目配置继承的方式

        allprojects {
              repositories {
                   jcenter()
              }
        }
}
```

图 1-39　项目的编译配置文件

repositories 属性主要用于配置库管理，默认支持 Java 依赖库；依赖属性中配置的是 Gradle 的版本；allprojects 属性中默认的是所有 Module 的编译配置继承 Project 的配置。

Module 的编译配置文件如图 1-40 所示；compileSdkVersion 属性代表 SDK 的编译版本；minSdkVersion 表示运行的最低要求；targetSdkVersion 表示运行的最高要求；dependencies 表示 Module 所需要的外部包的列表。

```
apply plugin: 'com.android.application' //构建为应用程序
android {
      compileSdkVersion 27       //SDK 编译版本
      defaultConfig {
            applicationId "cn.edu.sziit.a183_2_2"
            minSdkVersion 15          //运行最低要求
            targetSdkVersion 27       // 运行最高要求
            testInstrumentationRunner "android.support.test.runner.AndroidJUnitRunner " //测试
      }
      buildTypes {     //编译类型选择
            release {
                  minifyEnabled false
                  proguardFiles getDefaultProguardFile('proguard-android.txt'), 'proguard-rules.pro'
            }
      }
}
```

图 1-40　Module 的编译配置文件

2. 应用程序的运行过程

经过前面对 Android 项目目录结构和编译过程的介绍，我们对 Android 的文件结构已经

有所了解，下面介绍 Android 应用程序的运行过程。Android 程序执行的整个序列图如图 1-41 所示。

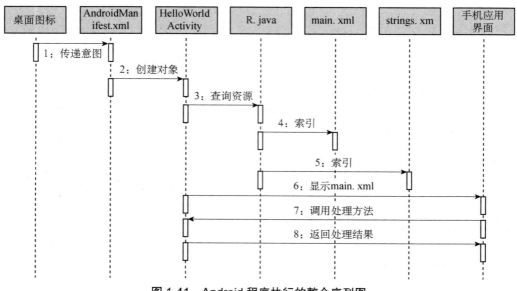

图 1-41 Android 程序执行的整个序列图

（1）发布程序到手机上之后，当双击手机桌面上该应用的图标时，系统会将这个单击事件包装成一个 Intent（意图），该 Intent 包含两个参数，如图 1-42 所示。

```
<activity android:name=".MainActivity">
    <intent-filter>
        <action android:name="android.intent.action.MAIN" />
        <category android:name="android.intent.category.LAUNCHER" />
    </intent-filter>
</activity>
```

图 1-42 AndroidManifest.xml 意图

（2）Intent 被传递给 HelloWorld 应用后，在应用的功能清单文件 AndroidManifest.xml 中寻找与该 Intent 匹配的 Intent 过滤器"<intent-filter>"，如果匹配成功，找到相匹配的 Intent 过滤器所在的 Activity 元素，再根据<activity>元素的"name"属性来寻找其对应的 Activity 类"com.sziit.soft.task2_helloworld.MainActivity"。

（3）Android 操作系统创建该 Activity 类的实例对象，对象创建完成之后，会执行到该类的 onCreate 方法，此 OnCreate 方法是重写其父类 Activity 的 OnCreate 方法而实现的。onCreate 方法用来初始化 Activity 实例对象。如图 1-43 所示的是"MainActivity.java"类中 onCreate 方法的代码。

```
public class MainActivity extends AppCompatActivity {
    @Override
    protected void onCreate(Bundle savedInstanceState) {
        super.onCreate(savedInstanceState);
        setContentView(R.layout.activity_main);
    }
}
```

图 1-43 MainActivity.java 类中 onCreate 方法的代码

其中 super.onCreate(savedInstanceState)的作用是调用其父类 Activity 的 OnCreate 方法来实现对界面的画图绘制工作。在实现自己定义的 Activity 子类的 OnCreate 方法时一定要记得调用该方法，以确保能够绘制界面。

setContentView(R.layout.activity_main)的作用是加载一个界面。该方法中传入的参数是"R.layout.main"，其含义为 R.java 类中静态内部类 layout 的静态常量 main 的值，而该值是一个指向 res 目录下的 layout 子目录下 activity_main.xml 文件的标识符，代表着显示 activity_main.xml 所定义的画面。

activity_main.xml 文件中定义的控件（比如文本显示控件）引用了 string.XML 的变量，需要将 string.XML 的变量的值取出来进行显示。

引用完成后，系统真正地通过 setContentView(R.layout. activity_main)将界面显示给用户。

（4）用户在手机界面上进行单击按钮和触摸屏幕的操作。

（5）Android 系统监控用户的行为，并对用户的行为进行处理和反馈。

关于 Activity 类的执行流程及其生命周期会在后面的部分详细讲解。

1.2.3　单元小测

单选题：

1. Android 的程序创建中，Application name 表示（　　）。
 A. 项目名称　　　　B. 项目包名　　　　C. 项目类名称　　　　D. 应用程序名称
2. Android 的程序创建中，Company domains 表示（　　）。
 A. 项目名称　　　　B. 项目包名　　　　C. 应用程序域名　　　　D. 应用程序名称
3. Android 的程序创建中，Package name 表示（　　）。
 A. 项目名称　　　　B. 项目包名　　　　C. 项目类名称　　　　D. 应用程序名称
4. Android 的程序创建中，会生成一个默认的 Activity 的名字为（　　）。
 A. MainActivity　　　B. EmptyActivity　　　C. Main　　　　D. Activity
5. Android 的程序创建中，每个 Activity 会生成一个默认的布局文件的名字为（　　）。
 A. activity　　　　B. main　　　　C. main_activity　　　　D. activity_main
6. Android 的程序创建中，布局文件采用的格式为（　　）。
 A. xml　　　　B. html　　　　C. java　　　　D. javascript
7. Android 的程序创建中，布局文件保存在项目的哪个文件夹中？（　　）
 A. java\main　　　B. res\values　　　C. res\drawable　　　D. res\layout
8. Android 的程序创建中，字符串文件保存在项目的哪个文件夹中？（　　）
 A. java\main　　　B. res\values　　　C. res\drawable　　　D. res\layout
9. Android 程序启动最先加载 AndroidManifest.xml 文件，如果有多个 Activity，请问以下哪个属性决定了 Activity 最先被加载？（　　）
 A. android.intent.action.ICON　　　　B. android.intent.action.LAUNCHER
 C. android.intent.action.Main　　　　D. android.intent.action.ICON
10. Android 应用程序的后缀名为（　　）。
 A. exe　　　　B. apk　　　　C. jar　　　　D. tar

Android 程序调试
（慕课）

1.3 Android 应用程序调试

1.3.1 Android 调试工具

1. DDMS

DDMS 的全称是 Dalvik Debug Monitor Service，是 Android 开发环境中的 Dalvik 虚拟机调试监控服务。DDMS 提供例如测试设备截屏，针对特定的进程查看正在运行的线程及堆信息、Logcat、广播状态信息、模拟电话呼叫、接收 SMS、虚拟地理坐标等。DDMS 存放在 SDK-tools 路径下，可以直接双击 ddms.bat 运行；也可以在 Android Studio 环境中通过"已运行的模拟器"进行启动，如图 1-44 所示；DDMS 对 Emulator 和外接测试机同等效用，如果系统检测到它们（VM）同时运行，用户需要自己选择需要监控的设备。

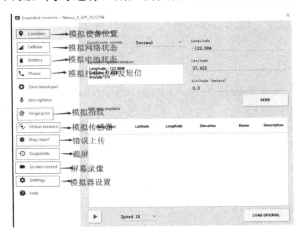

图 1-44 Android Studio DDMS 界面

Android Studio 开发环境的右下角是"Device File Explorer"设备文件查询工具，可以查看连接的模拟器或者手机的所有文件信息，如图 1-45 所示。

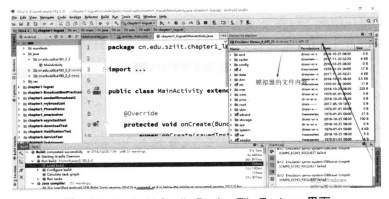

图 1-45 Android Studio Device File Explorer 界面

模拟器的功能主要包括：Location（模拟设备位置）、Cellular（模拟网络状态）、Battery（模拟电池状态）、Phone（模拟电话和短信）、Fingerprint（模拟指纹）、Virtual Sensors（模拟传感器）、Snapshots（模拟器截屏）、Screen record（屏幕录像）、Setting（模拟器设置）。

DDMS 将搭建起 IDE 与测试终端（Emulator 或者 connected device）的连接，它们应用各自独立的端口监听调试信息，DDMS 可以实时监测到测试终端的连接情况。当有新的测试终端连接后，DDMS 将捕捉到终端的 ID，并通过 ADB 建立调试器，从而实现发送指令到测试终端的目的。

DDMS 的工作原理如图 1-46 所示。App VMs 产生 Log 日志信息，并通过与 ADB Device Daemon 之间的交互输出日志信息，而 ADB Device Daemon 又通过相应的协议通过 USB（Device 手机设备）或本地连接（Emulator 模拟设备），与 PC 上运行的 ADB Host Daemon 交互，通过 PC 上的调试工具呈现给用户。

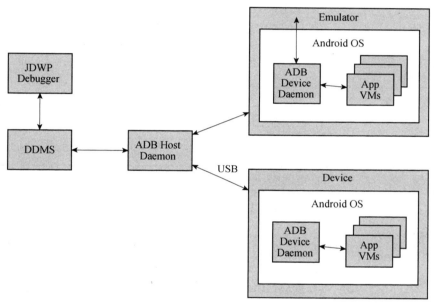

图 1-46　DDMS 的工作原理

2. Log 日志功能

Java 中我们都很喜欢使用传统的"System.out.println()"打印日志，不过在真正的项目开发中，很少会建议大家使用；System.out.println()除了使用方便，但是由于日志打印不受控制，打印时间和位置也不能确定，也不能添加过滤器，日志也没有级别的区分，因此在 Android Studio 中已不支持这种方法。

在 Android 应用程序的调试中一般使用 LogCat 信息窗口。LogCat 信息窗口中会显示应用程序的运行信息，包括调试信息、警告信息、错误信息、普通信息及冗余信息。不同类型的信息具有不同的显示颜色，方便开发人员观察。

在 DDMS 中的 LogCat 可以输出应用程序的运行信息，如果开发人员需要在程序运行时打印一些调试用的消息，可以使用 android.util 包下的 Log 类，该类可以将信息以日志的形式输出到 LogCat 中。

android.util.Log 常用的方法有以下 5 个：Log.v()、Log.d()、Log.i()、Log.w()及 Log.e()。

根据首字母对应 verbose、debug、information、warning、error。

Log.v 的调试颜色为黑色，任何消息都会输出，这里的 v 代表 verbose 琐碎（日志级别最小的意思），平时使用的就是 Log.v("","")。

Log.d 的输出颜色是蓝色，仅输出 debug（调试）的意思，但它会输出上层的信息，过滤后可以通过 DDMS 的 Logcat 标签来选择。

Log.i 的输出颜色为绿色，一般用于提示性的消息 information，它不会输出 Log.v 和 Log.d 的信息，但会显示 i、w 和 e 的信息。

Log.w 的输出颜色为橙色，可以看作 warning 警告，一般需要我们注意优化 Android 代码，同时选择它后还会输出 Log.e 的信息。

Log.e 的输出颜色为红色，可以想到 error 错误，这里仅显示红色的错误信息，这些错误就需要我们认真分析，查看栈的信息了。

Log 类中主要用到的方法及说明如表 1-1 所示。

表 1-1　Log 类中主要用到的方法及说明

方法名	方法说明	参数说明
Log.v(String tag,String msg)	输出冗余信息	tag：日志标签，可用于过滤日志信息 msg：输出的日志信息
Log.d(String tag,String msg)	输出调试信息	
Log.i(String tag,String msg)	输出普通信息	
Log.w(String tag,String msg)	输出警告信息	
Log.e(String tag,String msg)	输出错误信息	

Android 程序调试
（实践案例）

1.3.2　Android 调试实现

本小节的任务是使用 LogCat 对 Android 的应用程序进行监控，程序启动运行的时候能在开发环境的 DDMS 视图中输出 Android 程序的调试信息。完成本任务后，Android 程序运行后 DDMS 视图监控信息如图 1-47 所示。

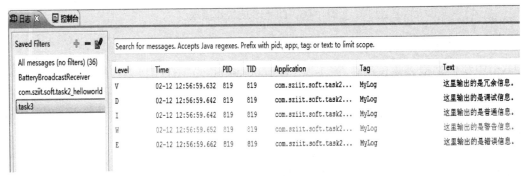

图 1-47　Android 程序运行后 DDMS 视图监控信息

1. 新建 Module

（1）打开 Android Studio 开发环境，在开发工具环境视图"菜单栏"中右击，在弹出的快捷菜单中选择"File"→"New"→"Create Module"命令，进入新建 Module 界面。

（2）在新建 Module 界面中选择"Phone & Tablet Module"，如图 1-48 所示；单击"Next"按钮进入 Module 设置界面。

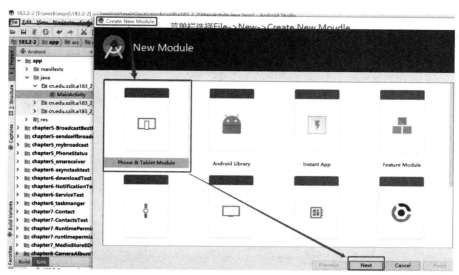

图 1-48　新建 Module 界面

（3）在 Module 设置界面中，需要设置 Application 的名字，也就是 Module 的名字，设置后单击"Next"按钮完成 Module 的新建，如图 1-49 所示。

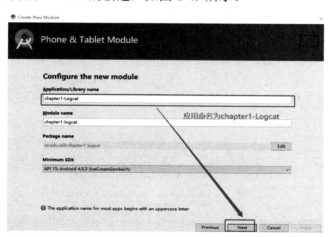

图 1-49　Module 设置界面

2. 调试代码编写

上一节中已经新建"chapter1-logcat"的 Module，本节需要对项目中的一些内容添加调试信息。

（1）在开发工具环境视图"包资源管理器"中单击"chapter1-logcat"项目的 Java 文件目录，打开"MainActivity.java"文件；在文件的开头添加"android.util.Log"库（Android 中负责输出调试信息）；类 android.util.Log 的构造函数是私有的，并不会被实例化，只是提供了静态的属性和方法。

（2）将调试信息添加到 MainActivity 类的"onCreate"函数中；要输出 Log 信息，可直接调用 Log.v()、Log.d()、Log.i()、Log.w()、Log.e()等类方法；如图 1-50 所示的加粗字体代码；在 MainActivity 类中新建一个 initData 函数；定义 String 变量 Tag 用于获取类名的全部信息，并使用 Log 依次输出冗余、调试、普通、警告和错误信息；程序运行后在 Android 的输出控制台使用 Logcat 日志抓取窗口，可以看到程序的运行日志调试信息；Log 的主要用处是程序员在调试程序中加入调试信息便于后期的程序维护。

```
package cn.edu.sziit.chapter1_logcat;
import android.support.v7.app.AppCompatActivity;
import android.os.Bundle;
import android.util.Log;
public class MainActivity extends AppCompatActivity {
    @Override
    protected void onCreate(Bundle savedInstanceState) {
        super.onCreate(savedInstanceState);
        setContentView(R.layout.activity_main);
        initData();
    }
    private void initData() {
        String TAG=getLocalClassName().toString();
        Log.d(TAG, "initData:调试信息 ");
        Log.v(TAG,"initData:冗余信息");
        Log.i(TAG, "initData: 普通信息");
        Log.w(TAG, "initData: 警告信息");
        Log.e(TAG, "initData: 错误信息");
    }
}
```

图 1-50　增加调试代码

3. 调试项目

（1）在开发工具环境视图的工具栏中，选中"chapter1-logcat"Module，选择手机模拟器设备"Nexus 4 API 28"中，单击右边的"运行"按钮，编写的应用程序可以运行到模拟器上，如图 1-51 所示。

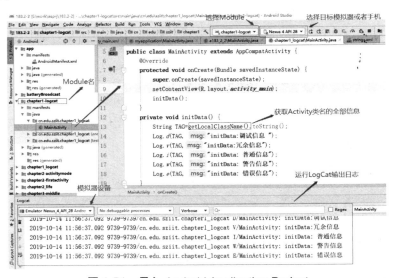

图 1-51　导入 Android Application Project

（2）在日志视图中选择"Edit Filter Configuration"选项，如图 1-52 所示，单击绿色的"+"号按钮进入"Create New Logcat Filter"（新建新日志过滤器）对话框。

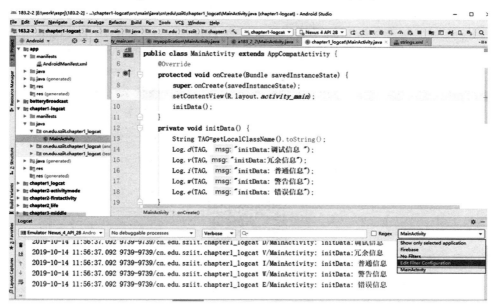

图 1-52　新建"Tag=" MainActivity""过滤器

（3）新建一个"Filter Name"过滤名为"MainActivity"的过滤器。图 1-50 中的代码"String TAG=getLocalClassName().toString()"，表示输出信息的 Tag 为"MainActivity 类名"，在"by Log Tag"选项中输入"空"；可以将图 1-51 所示的日志信息通过"Tag="MainActivity""进行过滤；填写完成后单击"确定"按钮；在日志视图的"Filters"选项中单击"MainActivity"，输入的日志只显示 Tag 为"MainActivity"的调试信息，如图 1-53 所示。

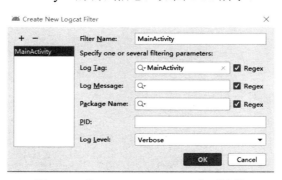

图 1-53　"Tag=" MainActivity ""过滤器

1.3.3　单元小测

单选题：

1. Android 的程序调试中，使用下面哪一个可以输出调试信息？（　　）
A. Log.e　　　　　　B. Log.d　　　　　　C. Log.i　　　　　　D. Log.w

2. Android 的程序调试中，使用下面哪一个可以输出普通提示信息？（   ）
A. Log.e   B. Log.d   C. Log.i   D. Log.w
3. Android 的程序调试中，使用下面哪一个可以输出警告信息？（   ）
A. Log.e   B. Log.d   C. Log.i   D. Log.w
4. Android 的程序调试中，使用下面哪一个可以输出错误信息？（   ）
A. Log.e   B. Log.d   C. Log.i   D. Log.w
5. Android 的程序调试中，getLocalClassName.toString 代表什么意思？（   ）
A. 获取当前程序名   B. 获取当前程序包名
C. 获取当前程序类名   D. 获取当前程序域名

本章课后练习和程序源代码

第 1 章源代码及课后习题

2 Android 视图

知识点

Android 视图的创建和跳转、Android 视图的生命周期、Android 控件的事件响应、Android 的启动模式。

能力点

1. 熟练掌握 Android 的控件事件响应使用方法。
2. 熟练掌握 Android 的视图创建和视图跳转方法。

2.1 Android 视图概述

2.1.1 Activity

Activity 概述（慕课）

1. 生命周期

Activity 称为"活动"，在应用程序中，一个活动（Activity）通常就是一个单独的屏幕。每一个活动都被实现为一个独立的类，并且从活动基类中继承而来，活动类将会显示由视图控件组成的用户接口，并对事件做出响应。大多数的应用都是由多个 Activity 显示组成的，例如，第一个屏幕是用户的登录界面，第二个屏幕用来显示登录成功后的界面。

这里的每一个屏幕就是一个 Activity（活动），很容易实现从一个屏幕到一个新的屏幕，并且完成新的活动。当一个新的屏幕打开后，前一个屏幕将会暂停，并保存在历史栈中。用户可以返回到历史栈中的前一个屏幕，当屏幕不再使用时，还可以从历史栈中删除。

简单地理解为，Activity 代表一个用户所能看到的屏幕，主要用于处理应用程序的整体性工作，例如，监听系统事件（按键事件、触摸屏滑动事件等），为用户显示指定的 View，启动其他 Activity 等。目前所有应用的 Activity 都继承 android.support.v7.app.AppCompatActivity 类，该类是由 Android 提供的基础类 Activity 发展而来的，其他的 Activity 继承父类后，通过父类的方法来实现各种功能。

Activity 的整个生命周期都定义在下面的接口方法中，所有方法都可以被重载。所有的 Activity 都需要实现 onCreate(Bundle)去初始化设置，大部分 Activity 需要实现 onPause()去提交更改过的数据，当前大部分的 Activity 也需要实现 onFreeze()接口，以便恢复在 onCreate(Bundle)里面设置的状态。Activity 的接口方法如图 2-1 所示。

```
public class Activity extends ApplicationContext {
    protected void onCreate(Bundle icicle);
    protected void onStart();
    protected void onRestart();
    protected void onResume();
    protected void onFreeze(Bundle outIcicle);
    protected void onPause();
    protected void onStop();
    protected void onDestroy();
}
```

图 2-1　Activity 的接口方法

onCreate：当活动第一次启动的时候，触发该方法，可以在此时完成活动的初始化工作。

onStart：该方法的触发表示所属活动将被展现给用户。

onResume：当一个活动和用户发生交互的时候，触发该方法。

onPause：当一个正在前台运行的活动因为其他的活动需要前台运行而转入后台运行的时候，触发该方法。这时候需要将活动的状态持久化，比如正在编辑的数据库记录等。

onStop：当一个活动不再需要展示给用户的时候，触发该方法。如果内存紧张，系统会直接结束这个活动，而不会触发 onStop 方法。所以保存状态信息应该在 onPause 时做，而不是在 onStop 时做。活动如果没有在前台运行都将被停止，或者 Linux 管理进程为了给新的活动预留足够的存储空间而随时结束这些活动。因此，对于开发者来说，在设计应用程序的时候，必须时刻牢记这一原则。在一些情况下，onPause 方法或许是活动触发的最后的方法，因此，开发者需要在这个时候保存需要保存的信息。

onRestart：当处于停止状态的活动需要再次展现给用户的时候，触发该方法。

onDestroy：当活动销毁的时候，触发该方法。与 onStop 方法一样，如果内存紧张，系统会直接结束这个活动而不会触发该方法。

图 2-2 所示的 Activity 生命周期是 Android 官方提供的图例，描述了 Activity 生命周期的整个过程。

可以看到，Activity 有 3 个关键的生命周期循环。

一个 Activity 完整的生命周期自第一次调用 onCreate(Bundle)开始，直至调用 onDestroy()为止。Activity 在 onCreate()中设置所有"全局"状态以完成初始化，而在 onDestroy()中释放所有系统资源。比如说，如果 Activity 有一个线程在后台运行以便从网络上下载数据，它会以 onCreate()创建那个线程，而以 onDestroy()销毁那个线程。

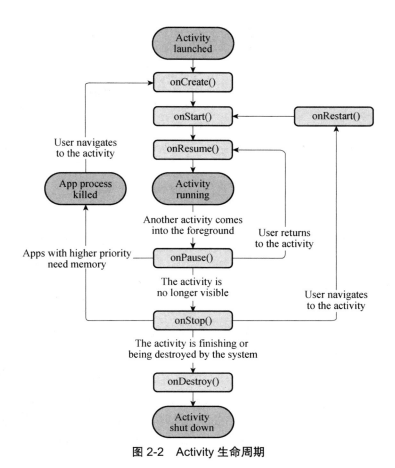

图 2-2 Activity 生命周期

一个 Activity 的可视生命周期自 onStart()调用开始直到相应的 onStop()调用为止。在此期间，用户可以在屏幕上看到此 Activity，尽管它也许并不是位于前台或者正在与用户做交互。在这两个方法中，你可以管控用来向用户显示这个 Activity 的资源。比如说，你可以在 onStart()中注册一个 BroadcastReceiver 来监控会影响到你 UI 的改变，而在 onStop()中来取消注册，这时用户是无法看到你的程序显示的内容的。onStart()和 onStop()方法可以随着应用程序是否为用户可见而被多次调用。

一个 Activity 的前台生命周期自 onResume()调用起，至相应的 onPause()调用为止。在此期间，Activity 位于前台的最上面并与用户进行交互。Activity 会经常在暂停和恢复之间进行状态转换，比如说当设备转入休眠状态或有新的 Activity 启动时，将调用 onPause()方法。当 Activity 获得结果或者接收到新的 Intent 时会调用 onResume()方法。因此，在这两个方法中的代码应当是轻量级的。

与生命周期相对应，Activity 主要有如下 4 种状态，4 种状态通过调用 Activity 的内部方法来进行转换，如图 2-3 所示。

Running（运行）：在屏幕前台（位于当前任务堆栈的顶部）。

Paused（暂停）：失去焦点但仍然对用户可见（覆盖 Activity 可能是透明或未完全遮挡的）。

Stopped（停止）：完全被另一个 Activity 覆盖。

Destroyed（销毁）：Activity 完全被销毁。

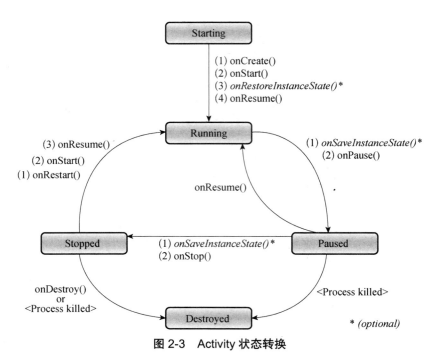

图 2-3 Activity 状态转换

一个 Activity 的实例的状态决定它在栈中的位置。处于前台的 Activity 总是在栈的顶端，当前台的 Activity 因为异常或其他原因被销毁时，处于栈第二层的 Activity 将被激活，上浮到栈顶。当新的 Activity 启动入栈时，原 Activity 会被压入到栈的第二层。一个 Activity 在栈中的位置变化反映了它在不同状态间的转换。Activity 的状态与它在栈中的位置关系如图 2-4 所示。

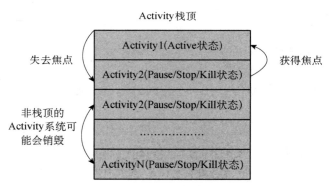

图 2-4 Andorid Activity 的状态与它在栈中的位置关系

除了顶层即处在 Active 状态的 Activity，其他的 Activity 都有可能在系统内存不足时被回收，一个 Activity 的实例越是处在栈的底层，它被系统回收的可能性越大。系统负责管理栈中 Activity 的实例，它根据 Activity 所处的状态来改变其在栈中的位置。

如果用户使用后退按钮返回的话，或者前台的 Activity 结束，活动的 Activity 就会被移出栈而消亡，而在栈上的下一个活动的 Activity 将会移上来并变为活动状态。

2. Activity 跳转

Intent 是 Android 程序中各组件之间进行交互的一种重要方式，它不仅可以指明当前组件想要执行的动作，还可以在不同组件之间传递数据。Intent 一般可被用于启动活动、启动服务及发送广播等场景，由于服务、广播等概念暂时还未涉及，那么我们的目光无疑就锁定在了启动活动上。Intent 大致可以分为两种：显式 Intent 和隐式 Intent。

（1）显式 Intent

Intent 有多个构造函数的重载，其中一个是 Intent(Context package Context, Class<?>cs)。这个构造函数接收两个参数，第一个参数 Context 要求提供一个启动活动的上下文，第二个参数 Class 则是指定想要启动的目标活动，通过这个构造函数就可以构建出 Intent 的"意图"。然后我们应该怎么使用这个 Intent 呢？Activity 类中提供了一个 startActivity()方法，这个方法是专门用于启动活动的，它接收一个 Intent 参数，这里我们将构建好的 Intent 传入 startActivity()方法就可以启动目标活动了。

在当前的 Module 中通过执行"File"→"New"→"Activity"→"Empty Activity"命令新建 SecondActivity 视图，如图 2-5 所示。

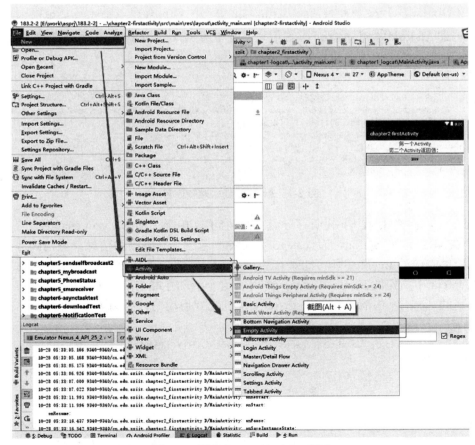

图 2-5　新建 SecondActivity 视图

从当前的 Activity 中启动另外一个 Activity 的显式启动的代码如图 2-6 所示。

```java
private void jumpToSecondActivity() {
    Intent mIntent=new Intent(MainActivity.this,SecondActivity.class);
    startActivity(mIntent);
}
```

图 2-6　Activity 显式跳转样例

我们首先构建出了一个 mIntent，传入 MainActivity.this 作为上下文，传入 SecondActivity.class 作为目标活动，这样我们的"意图"就非常明显了，即在 MainActivity.this 这个活动的基础上打开 SecondActivity 这个活动。然后通过 startActivity(mIntent)方法来执行这个 Intent，从而成功启动了 SecondActivity 这个活动。如果你想要回到上一个活动，怎么办呢？很简单，按下 Back 键就可以销毁当前活动，从而回到上一个活动了。Intent 的"意图"非常明显，因此我们称为显式 Intent。

（2）隐式 Intent

相对于显式 Intent，隐式 Intent 则含蓄了许多，它并不明确指出我们想要启动哪一个活动，而是指定了一系列更为抽象的 action 和 category 等信息，然后交由系统去分析这个 Intent 并帮助我们找出合适的活动去启动。那么 SecondActivity 可以响应什么样的隐式 Intent 呢？

通过在<activity>标签下配置<intent-filter>的内容，可以指定当前活动能够响应的 action 和 category，在 AndroidManifest.xml 配置信息文件中对 SecondActivity 进行配置，如图 2-7 所示。

```xml
<application>
<activity android:name=".SecondActivity">
    <intent-filter>
        //设置 SecondActivity 的隐式意图，只能设置一个
        <action android:name= "cn.edu.sziit.chapter2_firstactivity.ACTION_START"></action>
        //设置 SecondActivity 所在的目录，可以设置多个
        <category android:name="android.intent.category.DEFAULT"></category>
    </intent-filter>
</activity>
</application>
```

图 2-7　Activity 隐式跳转配置

在<action>标签中我们指明了当前活动可以响应 cn.edu.sziit.chapter2_firstactivity.ACTION_START 这个 action，而<category>标签则包含了一些附加信息，更精确地指明了当前的活动能够响应的 Intent 中还可能带有的 category；只有 action 和 category 中的内容同时能够匹配上 Intent 中指定的 action 和 category 时，这个活动才能响应该 Intent。

从当前的 Activity 中启动另外一个 Activity 的隐式启动的代码如图 2-8 所示。

```java
private void jumpToSecondActivity() {
    Intent mIntent=new Intent("cn.edu.sziit.chapter2_firstactivity.ACTION_START");
    startActivity(mIntent);
}
```

图 2-8　Activity 隐式跳转样例

可以看到，我们使用了 Intent 的另一个构造函数，直接将 action 的字符串传入，启动能够响应"cn.edu.sziit.chapter2_firstactivity.ACTION_START"这个 action 的活动；android.intent.category.DEFAULT 是一种默认的目录，用 startActivity()方法的时候会自动将这个 category 添加到 Intent 中。

2.1.2 Activity 生命周期实例

Activity 生命周期
（实践案例）

本小节以一个实例说明 Activity 生命周期。通过 Activity 生命周期的实例查看生命周期运行的全流程，如图 2-9 所示。

视图启动过程中会依次运行 onCreate->onStart->onResume。单击 Home 键，视图失去焦点，依次运行 onPause->onSaveInstanceState->onStop。重新运行视图，视图回到前台，依次运行 onRestart->onStart->onResume。关闭视图，依次运行 onPause->onSaveInstanceState->onStop->onDestroy。

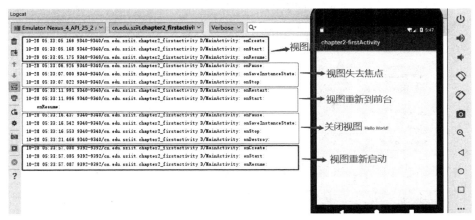

图 2-9 生命周期运行的全流程

1. 新建 Module

（1）打开 Android Studio 开发环境，在开发工具环境视图的菜单栏中右击，在弹出的快捷菜单中选择"File"→"New"→"Create Module"命令，进入新建 Module 界面。

（2）在新建 Module 界面中选择"Phone & Tablet Module"，如图 2-10 所示，单击"Next"按钮进入 Module 设置界面。

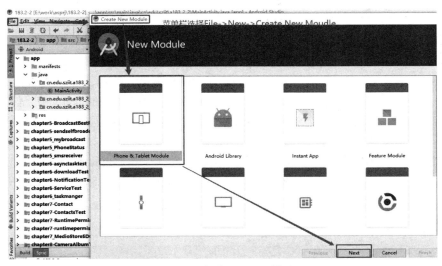

图 2-10 新建 Module 界面

（3）在 Module 设置界面中，需要设置 Application 的名字也就是 Module 的名字，单击"Next"按钮完成 Module 的新建，如图 2-11 所示。

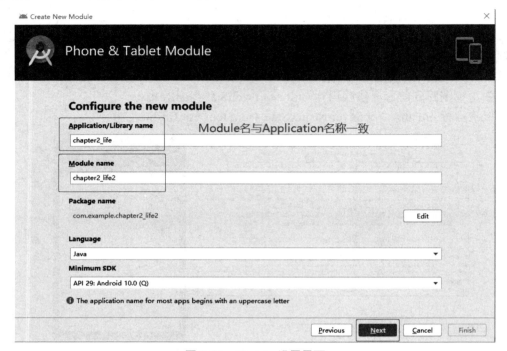

图 2-11　Module 设置界面

2. 代码编写

上一节中已经新建"chapter2-life"的 Module，本节需要对项目中的一些内容添加 Activity 生命周期代码。

（1）在开发工具环境视图"包资源管理器"中单击"chapter2-life"项目的 java 文件目录，打开"MainActivity.java"文件。

（2）使用 Ctrl+O 快捷键，可以依次重写 AppCompatActivity 的接口方法，如图 2-12 所示。

（3）在 MainActivity 类中定义一个私有的字符变量 TAG；在 MainActivity 中重写 onCreate 初始化方法；通过调用 getLocalClassName 系统函数将当前类的全名赋值给 TAG；在 onCreate 函数结尾使用 Log.d 打上标签；依次重写 onResume、onPause、onStart、onStop、onRestart、onDestroy，并在函数结尾使用 Log.d 打上标签，如图 2-13 所示。

图 2-12　"Ctrl+O"快捷键重写类方法

```
package com.example.chapter2_life;
import android.net.Uri;
import android.support.v7.app.AppCompatActivity;
import android.os.Bundle;
import android.util.Log;
```

图 2-13　重写父类方法

```java
public class MainActivity extends AppCompatActivity {
    private String TAG;
    @Override
    protected void onCreate(Bundle savedInstanceState) {
        super.onCreate(savedInstanceState);
        setContentView(R.layout.activity_main);
        TAG=(getPackageName()+"\\"+getLocalClassName()).toString();
        Log.d(TAG,"onCreate:    ");
    }
    //CTRL+O 快捷键重写父类方法
    @Override
    protected void onSaveInstanceState(Bundle outState) {
        super.onSaveInstanceState(outState);
        Log.d(TAG,"onSaveInstanceState:    ");
    }
    @Override
    protected void onRestoreInstanceState(Bundle savedInstanceState) {
        super.onRestoreInstanceState(savedInstanceState);
        Log.d(TAG,"onRestoreInstanceState:    ");
    }
    @Override
    protected void onRestart() {
        super.onRestart();
        Log.d(TAG,"onRestart:    ");
    }
    @Override
    protected void onStart() {
        super.onStart();
        Log.d(TAG,"onStart:    ");
    }
    @Override
    protected void onStop() {
        super.onStop();
        Log.d(TAG,"onStop:    ");
    }
    @Override
    protected void onDestroy() {
        super.onDestroy();
        Log.d(TAG,"onDestroy:    ");
    }
    @Override
    protected void onResume() {
        super.onResume();
        Log.d(TAG,"onResume:    ");
    }
    @Override
    protected void onPause() {
        super.onPause();
        Log.d(TAG,"onPause:    ");
    }
}
```

图 2-13　重写父类方法（续）

3. 运行项目

在开发工具环境视图的工具栏中，选中"chapter2-life" Module，选择手机模拟器设备"Nexus 4 API 28"中，单击右边的"运行"按钮，编写的应用程序可以运行到模拟器上，如图 2-9 所示。

Activity 跳转和
数据传递
（实践案例）

2.1.3 Activity 数据传递实例

本小节以一个实例介绍 Activity 的跳转和数据传递。

1. Activity 跳转

首先我们介绍 Activity 的跳转。在第一个 Activity 中单击"跳转到第二个 ACTIVITY"，页面跳转到第二个 Activity，如图 2-14 所示。

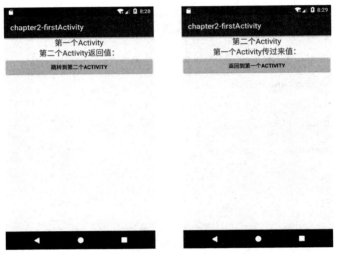

图 2-14　Activity 的跳转实例

（1）新建 Module

打开 Android Studio 开发环境，在开发工具环境视图的菜单栏中右击，在弹出的快捷菜单中选择"File"→"New"→"Create Module"命令，进入新建 Module 界面。

在新建 Module 界面中选择"Phone & Tablet Module"，如图 2-15 所示，单击"Next"按钮进入 Module 设置界面。

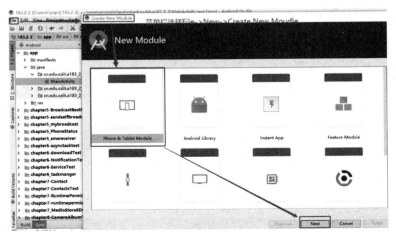

图 2-15　新建 Module 界面

在 Module 设置界面中，需要设置 Application 的名字也就是 Module 的名字，单击"Next"

按钮完成 Module 的新建，如图 2-16 所示。

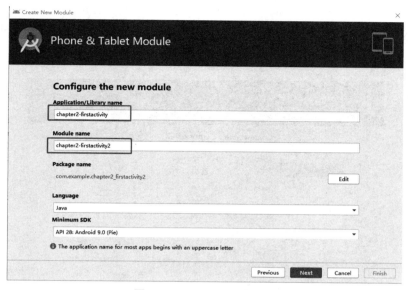

图 2-16 Module 设置界面

（2）新建视图

上一节中已经新建了"chapter2-firstactivity"的 Module，主页面的 MainActivity 视图类和 activity_main 布局文件已生成，如果要实现跳转，需要新建第二个页面的 SecondActivity 视图类和 activity_second 布局文件。

打开 Android Studio 开发环境，在开发工具环境视图的菜单栏中右击，在弹出的快捷菜单中选择"File"→"New"→"Activity"→"Empty Activity"命令，进入新建视图界面，如图 2-17 所示。

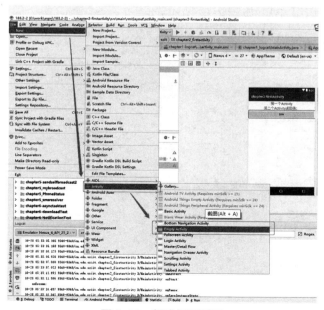

图 2-17 新建视图页面

在新建视图界面中，设置视图的名称为"SecondActivity"，选中"自动生成布局文件名"选项即"Generate Layout File"。Android Studio 自动为 Activity 生成一个布局文件的名称，如 activity_second，页面中其他的选项都采用默认设置，设置完成后，单击"Finish"按钮，新建视图完成，如图 2-18 所示。

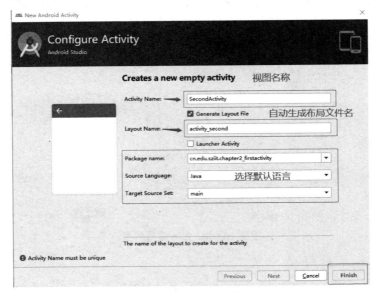

图 2-18　新建视图界面

（3）页面布局

在开发工具环境视图"包资源管理器"中单击"chapter2-firstactivity"项目的 res 文件目录，打开"activity_main.xml"布局文件。在布局文件中选择"Design"模式，布局文件使用线性布局，依次增加 textView1（标签组件）、editText（文本输入框组件）、textView2（标签组件）和 btn_jump（按钮组件），如图 2-19 所示。

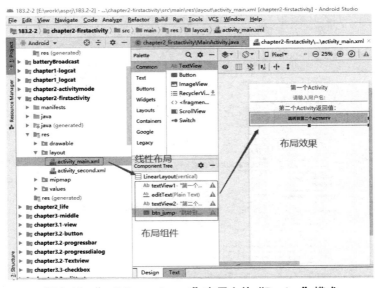

图 2-19　"activity_main.xml"布局文件"Design"模式

布局文件中选择"Text"模式，可以查看布局文件的 XML 语言代码，如图 2-20 所示。

```xml
<?xml version="1.0" encoding="utf-8"?>
<LinearLayout xmlns:android="http://schemas.android.com/apk/res/android"
    xmlns:app="http://schemas.android.com/apk/res-auto"
    xmlns:tools="http://schemas.android.com/tools"
    android:layout_width="match_parent"
    android:layout_height="match_parent"
    android:orientation="vertical"
    tools:context=".MainActivity">
    <TextView
        android:id="@+id/textView1"
        android:layout_width="match_parent"
        android:layout_height="wrap_content"
        android:gravity="center"
        android:text="第一个 Activity"
        android:textSize="20dp"
        app:layout_constraintBottom_toBottomOf="parent"
        app:layout_constraintLeft_toLeftOf="parent"
        app:layout_constraintRight_toRightOf="parent"
        app:layout_constraintTop_toTopOf="parent" />
    <EditText
        android:id="@+id/editText"
        android:layout_width="match_parent"
        android:layout_height="wrap_content"
        android:ems="10"
        android:gravity="center"
        android:hint="请输入用户名:"
        android:inputType="textPersonName" />
    <TextView
        android:id="@+id/textView2"
        android:layout_width="match_parent"
        android:layout_height="wrap_content"
        android:gravity="center"
        android:text="第二个 Activity 返回值："
        android:textSize="20dp"
        app:layout_constraintBottom_toBottomOf="parent"
        app:layout_constraintLeft_toLeftOf="parent"
        app:layout_constraintRight_toRightOf="parent"
        app:layout_constraintTop_toTopOf="parent" />
    <Button
        android:id="@+id/btn_jump"
        android:layout_width="match_parent"
        android:layout_height="wrap_content"
        android:layout_weight="0"
        android:text="跳转到第二个 Activity" />
</LinearLayout>
```

图 2-20 "activity_main.xml"布局文件的 XML 语言代码

在开发工具环境视图"包资源管理器"中单击"chapter2-firstactivity"项目的 res 文件目录，打开"activity_second.xml"布局文件。布局文件中选择"Design"模式，布局文件使用线性布局，依次增加 textView1（标签组件）、textView2（标签组件）和 btn_jump（按钮组件），如图 2-21 所示。

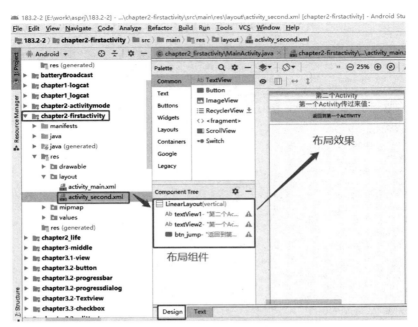

图 2-21 "activity_main.xml" 布局文件 "Design" 模式

布局文件中选择"Text"模式，可以查看布局文件的 XML 语言代码，如图 2-22 所示。

```xml
<?xml version="1.0" encoding="utf-8"?>
<LinearLayout xmlns:android="http://schemas.android.com/apk/res/android"
    xmlns:app="http://schemas.android.com/apk/res-auto"
    xmlns:tools="http://schemas.android.com/tools"
    android:layout_width="match_parent"
    android:layout_height="match_parent"
    android:orientation="vertical"
    tools:context=".MainActivity">
<TextView
    android:id="@+id/textView1"
    android:layout_width="match_parent"
    android:layout_height="wrap_content"
    android:gravity="center"
    android:text="第二个 Activity"
    android:textSize="20dp"
    app:layout_constraintBottom_toBottomOf="parent"
    app:layout_constraintLeft_toLeftOf="parent"
    app:layout_constraintRight_toRightOf="parent"
    app:layout_constraintTop_toTopOf="parent" />
<TextView
    android:id="@+id/textView2"
    android:layout_width="match_parent"
    android:layout_height="wrap_content"
    android:gravity="center"
    android:text="第一个 Activity 传过来值："
    android:textSize="20dp"
    app:layout_constraintBottom_toBottomOf="parent"
    app:layout_constraintLeft_toLeftOf="parent"
    app:layout_constraintRight_toRightOf="parent"
    app:layout_constraintTop_toTopOf="parent" />
<Button
    android:id="@+id/btn_jump"
    android:layout_width="match_parent"
```

图 2-22 "activity_second.xml" 布局文件的 XML 语言代码

```
            android:layout_height="wrap_content"
            android:layout_weight="0"
            android:text="返回到第一个 Activity" />
</LinearLayout>
```

图 2-22 "activity_second.xml" 布局文件的 XML 语言代码（续）

（4）编写跳转代码

LayoutCreator 是一款代码生成的插件，可以帮助提高 App 的开发速度，只要设计完成布局 XML 文件，可以让你在 Activity 和 Fragment 中自动生成 findViewById 等布局相关初始化代码。

在菜单栏中选择"File"→"Setting"→"Plugins"命令，安装"LayoutCreator"插件；也可以通过在线搜索安装和磁盘安装，如图 2-23 所示。

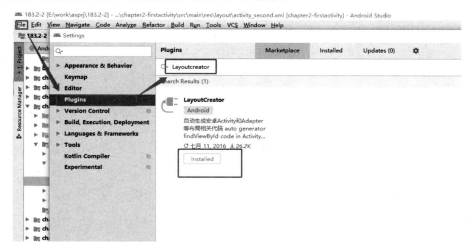

图 2-23 "LayoutCreator"插件安装界面

LayoutCreator 插件安装成功后，在开发工具环境视图"包资源管理器"中单击"chapter2-firstactivity"项目的 java 文件目录，打开"MainActivity.java"视图代码文件。在 onCreate 初始化函数中选择布局文件→右击 Generate→选择 LayoutCreator 插件。在打开的 LayoutCreator 插件界面中选择"mAaBbCc"驼峰命名法。自动对布局中的组件命名并初始化，如图 2-24 所示。

图 2-24 "LayoutCreator"自动对生成的组件命名并初始化代码

activity_main.java 组件初始化代码自动生成后，为跳转的 Button 组件增加事件响应代码，如图 2-25 所示。

增加一个 jumpToSecondActivity 方法实现第一个页面跳转回第二个页面,在按钮监听方法中调用 jumpToSecondActivity。MainActivity.java 的代码如图 2-25 所示。

```java
package cn.edu.sziit.chapter2_firstactivity;
import android.content.Intent;
import android.os.Bundle;
import android.support.v7.app.AppCompatActivity;
import android.util.Log;
import android.view.View;
import android.widget.Button;
import android.widget.EditText;
import android.widget.TextView;
public class MainActivity extends AppCompatActivity implements View.OnClickListener {
    private String TAG;
    private TextView mTextView1;
    private TextView mTextView2;
    private Button mBtnJump;
    private EditText mEditText;
    @Override
    protected void onCreate(Bundle savedInstanceState) {
        super.onCreate(savedInstanceState);
        setContentView(R.layout.activity_main);
        initView();
        initData();
    }
    //数据初始化代码
    private void initData() {
        TAG = getLocalClassName().toString();
        Log.d(TAG, "onCreate: ");
    }
    //组件初始化代码
    private void initView() {
        mTextView1 = (TextView) findViewById(R.id.textView1);
        mTextView2 = (TextView) findViewById(R.id.textView2);
        mBtnJump = (Button) findViewById(R.id.btn_jump);
        mBtnJump.setOnClickListener(this);
        mEditText = (EditText) findViewById(R.id.editText);
        mEditText.setOnClickListener(this);
    }
    //按钮响应代码
    public void onClick(View v) {
        switch (v.getId()) {
            case R.id.btn_jump:
                jumpToSecondActivity();
                break;
        }
    }
    //跳转代码
    private void jumpToSecondActivity() {
        Intent mIntent = new Intent(MainActivity.this, SecondActivity.class);
        startActivity(mIntent);
    }
}
```

图 2-25　MainActivity.java 代码

在开发工具环境视图"包资源管理器"中单击"chapter2-firstactivity"项目的 java 文件目录,打开"SecondActivity.java"视图代码文件;在 onCreate 初始化函数中选择布局文件→右击 Generate→选择 LayoutCreator 插件;在 LayoutCreator 插件界面中选择"mAaBbCc"驼峰命名法;自动对布局中的组件命名并初始化。

组件初始化完成后,增加一个 jumpToFirstActivity 方法,实现第二个页面跳转回第一个页面,在按钮监听方法和按键返回方法中分别调用 jumpToFirstActivity。SecondActivity.java 的代码，如图 2-26 所示。

```java
package cn.edu.sziit.chapter2_firstactivity;
import android.content.Intent;
import android.os.Bundle;
import android.support.v7.app.AppCompatActivity;
import android.view.View;
import android.widget.Button;
import android.widget.TextView;
public class SecondActivity extends AppCompatActivity implements View.OnClickListener {
    private TextView mTextView1;
    private TextView mTextView2;
    private Button mBtnJump;
    @Override
    protected void onCreate(Bundle savedInstanceState) {
        super.onCreate(savedInstanceState);
        setContentView(R.layout.activity_second);
        initView();
    }
//组件初始化代码
    private void initView() {
        mTextView1 = (TextView) findViewById(R.id.textView1);
        mTextView2 = (TextView) findViewById(R.id.textView2);
        mBtnJump = (Button) findViewById(R.id.btn_jump);
        mBtnJump.setOnClickListener(this);
    }
//按钮响应代码
    public void onClick(View v) {
        switch (v.getId()) {
            case R.id.btn_jump:
                jumpToFirstActivity();
                break;
        }
    }
//跳转到第一个视图
    private void jumpToFirstActivity() {
        Intent mIntent=new Intent(SecondActivity.this, MainActivity.class);
startActivity(mIntent);
    }
//返回键响应函数
    public void onBackPressed() {
        super.onBackPressed();
        jumpToFirstActivity();
    }
}
```

图 2-26　SecondActivity.java 的代码

2. Activity 数据传递

上一个小节中我们用 Intent 来启动一个活动，其实 Intent 还可以在启动活动的时候传递数据。在启动活动时传递数据的思路很简单，Intent 中提供了一系列 putExtra() 方法的重载，可以把想要传递的数据暂存在 Intent 中，启动了另一个活动后，只需要把这些数据再从 Intent 中取出就可以了。在 MainActivity 中将一个 "admin" 的字符串数据传递给 SecondActivity，MainActivity 中的 jumpToSecondActivity 方法代码如图 2-27 所示。

```java
public static final int REQUEST_CODE = 1;
//跳转到第一个视图
    private void jumpToSecondActivity() {
        String strData = "admin";
        Intent mIntent = new Intent(MainActivity.this, SecondActivity.class);
```

图 2-27　jumpToSecondActivity 方法代码

```
                mIntent.putExtra("user", strData);
                startActivityForResult(mIntent, REQUEST_CODE);
            }
```

图 2-27   jumpToSecondActivity 方法代码（续）

使用显式 Intent 的方式来启动 SecondActivity，并通过 putExtra()方法传递了一个字符串。putExtra()方法接收两个参数，第一个参数是键，用于后面从 Intent 中取值，第二个参数才是真正要传递的数据，然后我们在 SecondActivity 中将传递的数据取出并显示到文本组件中，如图 2-28 所示。

```
public class SecondActivity extends AppCompatActivity implements View.OnClickListener {
    @Override
    protected void onCreate(Bundle savedInstanceState) {
        super.onCreate(savedInstanceState);
        setContentView(R.layout.activity_second);
        initView();
        initData();
    }
    private void initData() {
        Intent mIntent = getIntent();
        String strPara = mIntent.getStringExtra("user");
        mTextView2.setText("第一个 Activity 传递的参数 user:"+strPara);
    }
```

图 2-28   SecondActivity 中接收传递数据

首先可以通过 getIntent()方法获取到用于启动 SecondActivity 的 Intent，然后使用 getStringExtra()方法传入相应的键值，就可以得到传递的数据。由于我们传递的是字符串，所以使用 getStringExtra()方法来获取传递的数据。如果传递的是整型数据，则使用 getIntExtra()方法；如果传递的是布尔型数据，则使用 getBooleanExtra()方法。

既然可以传递数据给下一个 Activity，那么能不能返回数据给上一个 Activity 呢？上面我们通过 startActivityForResult(mIntent,1)启动了 SecondActivity，REQUEST_CODE=1 这个参数是请求码，用于 SecondActivity 返回数据回调后判断数据的来源。SecondActivity 中可以重写 jumpToFirstActivity 方法，添加返回数据的功能，如图 2-29 所示。

```
//跳转到第一个视图
private void jumpToFirstActivity() {
    Intent mIntent=new Intent();
    mIntent.putExtra("password","sziit");
    setResult(RESULT_OK,mIntent);
    finish();
}
```

图 2-29   jumpToFirstActivity 中返回数据

可以看到，我们构建了一个 Intent 对象，这个 Intent 对象仅用于传递数据而已，它没有指定任何的"意图"。紧接着把要传递的数据存放在 Intent 对象中，然后调用 setResult 方法。setResult 专门用于向上一个活动返回数据。setResult 方法接收两个参数，第一个参数用于向上一个活动返回处理结果，一般只使用 **RESULT_OK** 或 **RESULT_CANCELED** 这两个值，第二个参数则把带有数据的 Intent 传递回去，然后调用 finish()方法来销毁当前活动。

由于使用了 startActivityForResult()方法来启动 SecondActivity，在 SecondActivity 被销毁之后会回调上一个活动的 onActivityResult()方法，因此我们需要在 MainActivity 中重写这个方法来得到返回的数据，如图 2-30 所示。

```
//SecondActivity 数据返回回调
protected void onActivityResult(int requestCode, int resultCode, Intent data) {
    super.onActivityResult(requestCode, resultCode, data);
    switch (requestCode) {
        case REQUEST_CODE:
            if (resultCode == RESULT_OK) {
                String strReturnData;
                strReturnData = data.getStringExtra("password");
                mTextView2.setText("SecondActivity 返回：" + strReturnData);
            }
            break;
        default:
    }
}
```

图 2-30　MainActivity 中重写 onActivityResult()方法

onActivityResult()方法带有三个参数，第一个参数 requestCode，即我们在启动活动时传入的请求码；第二个参数 resultCode，即我们在返回数据时传入的处理结果；第三个参数 data，带着返回数据的 Inten 对象。由于在一个活动中有可能调用 startActivityforResult()方法去启动很多不同的活动，每一个活动返回的数据都会回调到 onActivityResult()这个方法，因此我们首先要做的就是通过检查 requestCode 的值来判断数据来源。确定数据是从 SecondActivity 返回的之后，我们再通过 resultCode 的值来判断处理结果是否成功，最后从 Intent 对象取出值并显示到 mTextView 2 文本组件中。

代码编写完成后，运行程序的效果如图 2-31 所示；第一个 Activity 中输入用户名 admin；并将用户名传递给第二个 Activity，第二个 Activity 读取到第一个 Activity 传递的用户名 admin；第二个 Activity 跳转回第一个 Activity，并传回参数值 sziit；第一个 Activity 读取并显示参数值。

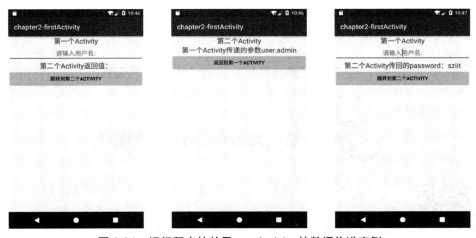

图 2-31　运行程序的效果——Activity 的数据传递实例

2.1.4　单元小测

选择题：

1. 一个 Android 应用程序默认会启动多少个 Activity？（　　）
A. 2　　　　　　　　B. 1　　　　　　　　C. 5　　　　　　　　D. 4
2. Activity 类中最先启动的是哪一个方法？（　　）

A. onCreate()　　　　B. onStart()　　　　C. onResume()　　　　D. onPause()

3. Activity 处于栈顶的时候处于什么状态？（　　）

A. 运行　　　　B. 暂停　　　　C. 停止　　　　D. 销毁

4. Activity 被某个 AlertDialog 遮住时处于什么状态？（　　）

A. 运行　　　　B. 暂停　　　　C. 停止　　　　D. 销毁

5. 请问下面的跳转方法属于什么跳转？（　　）

```
Intent mIntent=new Intent("cn.edu.sziit.chapter2_firstactivity.ACTION_START");startActivity (mIntent);
```

A. 显式　　　　B. 类跳转　　　　C. 配置加载　　　　D. 隐式

6. 请问下面的跳转方法属于什么跳转？（　　）

```
private void jumpToSecondActivity() {
Intent mIntent=new Intent(MainActivity.this,SecondActivity.class);
startActivity(mIntent);
}
```

A. 显式　　　　B. 类跳转　　　　C. 配置加载　　　　D. 隐式

7. 请问下面的配置的作用是什么？（　　）

```
<activity android:name=".SecondActivity"
android:theme="@style/Theme.AppCompat.Dialog"> </activity>
```

A. 配置 Activity 为显式启动　　　　B. 配置 Activity 为对话框启动
C. 配置 Activity 为隐式启动　　　　D. 配置 Activity 为正常布局启动

8. Activity 类中从启动到运行执行的方法是（　　）。（多选）

A. onCreate()　　　　B. onPause()　　　　C. onResume()　　　　D. onStart()

9. Activity 类中从暂停态到运行态执行的方法是（　　）。（多选）

A. onRestart()　　　　B. onPause()　　　　C. onResume()　　　　D. onStart()

10 Activity 类中从运行态到销毁态执行的方法是（　　）。（多选）

A. onStop()　　　　B. onPause()　　　　C. onDestroy()　　　　D. onStart()

2.2　Android 启动模式

Activity 应用（慕课）

Activity 4 种启动模式
（实践案例）

一个项目一般会包括多个 Activity，系统中使用任务栈来存储创建的 Activity 实例，任务栈是一种"后进先出"的栈结构。举个例子，若我们多次启动同一个 Activity，系统会创建多个实例并依次放入任务栈中。当按 Back 键返回时，每按一次，一个 Activity 出栈，直到栈空为止。当栈中没有 Activity 的时候，系统就会回收此任务栈。

上面这个样例中的 Activity 并没有设置启动模式，你会发现多次启动同一个 Activity，而系统却创建了多个实例，白白浪费了内存。Android 为 Activity 的创建提供了 4 种启动模式，依据实际应用场景的不同，为 Activity 选择不同的启动模式，最大化地降低内存的使用，因为每次都要在栈中创建一个新的 Activity。

Android 启动模式分为 4 种，分别是 standard、singleTop、singleTask 和 singleInstance。可以通过在 AndroidManifest.xml 中配置<activity>标签指定 android:launchMode 属性来选择启动模式。下面我们学习 Android 的 4 种启动模式。

2.2.1 standard

standard（标准模式）是活动默认的启动模式，如果没有为 Activity 设置启动模式，系统会默认它为标准模式，所有活动都会自动使用这种启动模式。目前为止我们写过的所有活动使用的都是 standard 模式。经过上一节的学习你已经知道了 Android 是使用返回栈来管理活动的，在 standard 模式下，每当启动一个新的活动，它就会在返回栈中入栈，并处于栈顶的位置。对于使用 standard 模式的活动系统不会在乎这个活动是否已经在返回栈中存在，每次启动都会创建该活动的一个新的实例。standard 模式在配置信息文件的配置信息如图 2-32 所示。

```
<activity android:name=".MainActivity"
    android:launchMode="standard">
    <intent-filter>
        <action android:name="android.intent.action.MAIN" />
        <category android:name="android.intent.category.LAUNCHER" />
    </intent-filter>
</activity>
```

图 2-32 standard 模式配置信息

standard 模式的栈内启动情况如图 2-33 所示。此时 Activity 栈中有 A、B、C 三个 Activity，C Activity 处于栈顶，启动模式为 standard 模式。若在 C Activity 中加入单击事件，要跳转到一个同类型的 C Activity，结果是新建一个 C Activity 进入栈中，成为栈顶。

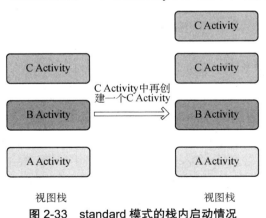

图 2-33 standard 模式的栈内启动情况

2.2.2 singleTop

在有些情况下 standard 模式不太合理，比如有的活动已经在栈顶了，再次启动的时候还需要创建一个对象实例呢？这种系统默认的一种启动模式不合理，Android 提供了 singleTop 启动模式（单栈顶模式）。当活动的启动模式指定为 singleTop 时，在启动活动时视图栈的栈顶已经是该活动,则认为可以直接使用它,不需要再创建新的活动实例;若需要创建的 Activity 未处于栈顶,此时会又一次创建一个新的 Activity 入栈,同 standard 模式一样。singleTop 模式在文件的配置信息如图 2-34 所示。

```
<activity android:name=".MainActivity"
    android:launchMode="singleTop">
    <intent-filter>
        <action android:name="android.intent.action.MAIN" />
        <category android:name="android.intent.category.LAUNCHER" />
    </intent-filter>
</activity>
```

图 2-34 singleTop 配置信息

singleTop 模式的栈内启动情况如图 2-35 所示。此时 Activity 栈中有 A、B、C 三个 Activity,C Activity 处于栈顶,启动模式为 singleTop 模式。若在 C Activity 中加入单击事件,要跳转到一个同类型的 C Activity,结果是直接复用栈顶的 C Activity。若在 C Activity 中加入单击事件,要跳转到不同类型的 A Activity,其结果是新建一个 C Activity 进入栈中,成为栈顶。

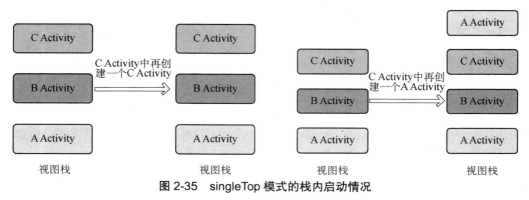

图 2-35 singleTop 模式的栈内启动情况

2.2.3 singleTask

用 singleTop 模式可以很好地解决重复创建栈顶活动的问题,但是如果该活动并没有处于栈顶的位置,还是可能会创建多个活动实例的。那么有什么办法可以让某个活动在整个应用程序的上下文中只存在一个实例呢?这就要借助 singleTask 模式来实现了。当活动的启动模式指定为 singleTask(单任务模式)时,每次启动该活动系统首先会在返回栈中检查是否存在该活动的实例,如果发现已经存在则直接使用该实例,并把在这个活动之上的所有活动统统出栈,如果没有发现就会创建一个新的活动实例。singleTask 模式在文件中的配置信息如图 2-36 所示。

```
<activity android:name=".MainActivity"
    android:launchMode=" singleTask ">
    <intent-filter>
        <action android:name="android.intent.action.MAIN" />
        <category android:name="android.intent.category.LAUNCHER" />
    </intent-filter>
</activity>
```

图 2-36 singleTask 配置信息

要创建的 Activity 已经处于栈中时,此时不会创建新的 Activity,而是将存在栈中的 Activity 上面的其他 Activity 所有销毁,使它成为栈顶。singleTask 模式的栈内启动情况如图 2-37 所示。此时 Activity 栈中有 A、B、C 三个 Activity,C Activity 处于栈顶,启动模式为 singleTask 模式。若在 C Activity 中加入单击事件,要跳转到一个同类型的 C Activity,其结果是直接复用栈顶的 C Activity。若在 C Activity 中加入单击事件,跳转到不同类型的 A Activity,其结果是将 A Activity 上面的 B Activity、C Activity 所有销毁,使 A Activity 成为栈顶。

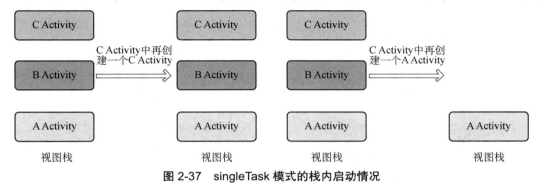

图 2-37　singleTask 模式的栈内启动情况

2.2.4　singleInstance

singleInstance 模式(单实例模式)不同于以上 3 种启动模式,指定为 singleInstance 模式的视图 Activity 会启用一个新的返回栈来管理这个活动。想象以下场景,假设系统中 Launch、锁屏键的应用程序中的活动是允许其他程序调用的,如果我们想要实现其他程序都可以共享这个活动的实例,使用前面 3 种启动模式肯定是做不到的,因为每个应用程序都会有自己的返回栈,同一个视图在不同的返回栈中入栈时必然会创建新的实例。而使用 singleInstance 模式式就可以解决这个问题,这种模式下会有一个单独的返回栈来管理这个活动,不管是哪个应用程序来访问这个活动,共用的都是同一个返回栈,也就解决了共享活动实例的问题。singleInstance 模式在文件中的配置信息如图 2-38 所示。

```
<activity android:name=".MainActivity"
    android:launchMode="singleInstance">
    <intent-filter>
        <action android:name="android.intent.action.MAIN" />
        <category android:name="android.intent.category.LAUNCHER" />
    </intent-filter>
</activity>
```

图 2-38　singleInstance 配置信息

打个比方,锁屏键应用 Activity 采用的是 singleInstance 模式,启动锁屏键应用后,系统会为它创建一个单独的任务栈。如果 A、B、C 三个应用都同时调用了锁屏键这个应用,当 B 应用退出的时候,锁屏键应用视图由于在一个单独的任务栈中,并不会退出,A、C 两个应用还可以继续调用锁屏键这个应用,如图 2-39 所示。

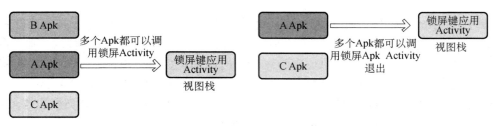

图 2-39　singleTask 视图栈

2.2.5　单元小测

选择题：

1. 请阅读下列代码：

```
private void jumpToSecondActivity() {
String strData = "admin";
Intent mIntent = new Intent(MainActivity.this, SecondActivity.class);
mIntent.putExtra(""user"", strData);
startActivityForResult(mIntent, REQUEST_CODE)
}
```

其中 startActivityForResult 方法的第二个参数的含义是什么？（　　）

A. 传送数据的标识符

B. 请求码，用于识别第二个 Activity 传递回来的数据

C. 结果标识符

D. 传递的数据

2. 请阅读下列代码：

```
protected void onActivityResult(int requestCode, int resultCode, Intent data) {
switch (requestCode) {
case REQUEST_CODE://标志符
if (resultCode == RESULT_OK) {
    strReturnData = data.getStringExtra(""password"");
      mTextView2.setText(""第二个 Activity 传回的 password："" + strReturnData);
    }
    break;
    default:
}
}
```

其中，onActivityResult 的作用是什么？（　　）

A. 处理第一个 Activity 的运行数据

B. 处理第一个 Activity 传递给第二个 Activity 的运行数据

C. 处理第二个 Activity 的运行数据

D. 处理第二个 Activity 传回给第一个 Activity 的运行数据

3. 请阅读下列代码：

```
private String strPara; private void initData() {
Intent mIntent = getIntent();
strPara = mIntent.getStringExtra(""user"");
```

```
mTextView2.setText(""第一个 Activity 传递的参数 user:""+strPara);
}
```

initData()函数的作用是什么？（   ）

A. 处理第一个 Activity 的运行数据
B. 处理第一个 Activity 传递给第二个 Activity 的运行数据
C. 处理第二个 Activity 的运行数据
D. 处理第二个 Activity 传回给第一个 Activity 的运行数据

4. 请阅读下列代码：

```
private void jumpToFirstActivity() {
    Intent mIntent=new Intent();
    mIntent.putExtra(""password"",""sziit"");
    setResult(RESULT_OK,mIntent);
    finish();
}
```

jumpToFirstActivity()函数的作用是什么？（   ）

A. 处理第一个 Activity 的运行数据
B. 处理第一个 Activity 传递给第二个 Activity 的运行数据
C. 处理第二个 Activity 的运行数据
D. 第二个 Activity 传回给第一个 Activity 的运行数据

5. 请阅读下列代码：

```
protected void onSaveInstanceState(Bundle outState) {
        super.onSaveInstanceState(outState);
        Log.d(TAG, ""onSaveInstanceState: "");
        outState.putString(""user"", mEditText.getText().toString());
    }
```

该代码可以完成的功能为（   ）。

A. 暂停时候保存数据   B. 重新启动时候保存数据
C. 视图销毁时保存数据   D. 启动时保存数据

6. 请阅读下列代码：

```
if (savedInstanceState != null) {
    String strUser = savedInstanceState.getString("user");
     mEditText.setText(strUser);
}
```

该代码可以完成的功能为（   ）。

A. 暂停时读取数据   B. 重新启动时读取数据
C. 视图销毁时读取数据   D. 启动时读取保存数据

7. 请问下列 Acvitiy 的配置采用了什么模式？（   ）

```
<activity android:name=".MainActivity"
android:launchMode="standard">
</activity>
```

A. 单任务模式   B. 标准模式   C. 单实例模式   D. 单栈顶模式

8. 请问下列 Acvitiy 的配置采用了什么模式？（   ）

```
<activity android:name=".MainActivity"
 android:launchMode="singleTop">
</activity>
```

A. 单任务模式　　　B. 标准模式　　　　C. 单实例模式　　　D. 单栈顶模式

9. 请问下列 Acvitiy 的配置采用了什么模式？（　　）

```
<activity android:name=".MainActivity"
 android:launchMode="singleTask">
 </activity>
```

A. 单任务模式　　　B. 标准模式　　　　C. 单实例模式　　　D. 单栈顶模式

10. 请问下列 Acvitiy 的配置采用了什么模式？（　　）

```
<activity android:name=".MainActivity"
 android:launchMode="singleInstance">
</activity>
```

A. 单任务模式　　　B. 标准模式　　　　C. 单实例模式　　　D. 单栈顶模式

本章课后练习和程序源代码

第 2 章源代码及课后习题

3 Android 布局与组件

 知识点

Android 的常用布局、Android 的组件、Android 适配器。

 能力点

1. 熟练掌握 Android 的各种布局方法。
2. 熟练使用 Android 的组件。
3. 熟练使用 Android 的适配器。

 3.1 Android 布局

Activity 布局
（慕课）

　　Android 的界面需要由多个控件来组成，如何才能让各个控件按照客户的需求进行摆放？这就需要使用布局。布局是一种可以放置很多控件的容器，它可以将控件按照一定的规律进行摆放，从而使界面美观大方。布局的内部还可以放置布局，通过多层布局的嵌套，可以完成一些复杂的界面，下面介绍 Android 的布局。

　　Android 的绝大部分 UI 组件都放在 android.widget 包及其子包 android.view 中，Android 应用的所有 UI 组件都继承了 View 类，View 组件非常类似于 Swing 变成的 JPanel，它代表一个空白的矩形区域。View 类还有一个重要的子类：ViewGroup，但 ViewGroup 通常作为其他组件的容器使用。Android 的所有 UI 组件都是建立在 View、ViewGroup 基础之上的，Android 采用"组合器"设计模式来设计 View 和 ViewGroup。ViewGroup 是 View 的子类，因此 ViewGroup 也可被当成 View 使用，对于一个 Android 应用的图形用户界面来说，ViewGroup 作为容器来盛装其他组件，而 ViewGroup 里除了可以包含普通 View 组件，还可以再次包含 ViewGroup 组件。

　　Android 提供了两种方式来控制组件的行为：在 XML 布局文件中通过 XML 属性进行控制；在 Java 代码中通过调用方法进行控制。实际上不管使用哪种方式，它们控制 Android 用户界面行为的本质是完全一样的。对于 View 类而言，它是所有 UI 组件的基类，因此它包含的 XML 属性和方法是所有组件都可以使用的。表 3-1 是 View 组件的 XML 属性列表。

表 3-1 View 组件的 XML 属性列表

属性名称	属性定义		
android:id	视图组件的编号：@+id/btn_jump		
android:layout_width	视图的宽度	dp	具体的 dp 数值
		match_parent	与上级视图一样宽
		wrap_content	与组件内容一样宽
android:layout_height	视图的高度	dp	具体的 dp 数值
		match_parent	与上级视图一样高
		wrap_content	与组件内容一样宽
android:layout_margin	视图与周围视图的距离	layout_marginTop	与上边视图距离
		layout_marginBottom	与下边视图距离
		layout_marginLeft	与左边视图距离
		layout_marginRight	与右边视图距离
android:background	视图的背景：可以是颜色，也可以是图片		
android:layout_gravity	视图与上级视图的对齐方式	left	靠左对齐
		right	靠右对齐
		top	靠上对齐
		bottom	靠下对齐
		center	居中对齐
android:gravity	视图中文字与视图的对齐方式	left	靠左对齐
		right	靠右对齐
		top	靠上对齐
		bottom	靠下对齐
		center	居中对齐
android:visible	视图的可视属性	visible	可见
		invisible	不可见，占位置
		gone	消失，不占位置
android:padding	内部内容与视图的距离	paddingTop	视图边缘与上边距离
		paddingBottom	视图边缘与下边距离
		paddingLeft	视图边缘与左边距离
		paddingRight	视图边缘与右边距离
app:layout_constraint	布局限制	app:layout_constraintBottom_toBottomOf	限制视图与底部对齐
		app:layout_constraintEnd_toEndOf	视图边缘与右边对齐
		app:layout_constraintStart_toStartOf	限制视图与左边对齐
		app:app:layout_constraintTop_toTopOf	限制视图与顶部对齐

android:id：视图组件的编号。

android:layout_width：视图的宽度；match_parent 代表与上级视图一样宽，wrap_content 代表与组件内容一样宽。

android:layout_height：视图的高度；match_parent 代表与上级视图一样高，wrap_content 代表与组件内容一样高。

android:layout_margin：视图与周围视图的距离。
- layout_marginTop 代表与上边视图距离。
- layout_marginBottom 代表与下边视图距离。
- layout_marginLeft 代表与左边视图距离。
- layout_marginRight 代表与右边视图距离。
- android:layout_minWidth：视图最小宽度。
- android:layout_minHeight：视图最小高度。

android:background：视图的背景，可以是颜色；也可以是图片。

android:layout_gravity：视图与上级视图的对齐方式；left 代表靠左对齐，right 代表靠右对齐，top 代表靠上对齐；bottom 代表靠下对齐，center 代表居中对齐。

android:gravity：视图中文字与视图的对齐方式；left 代表靠左对齐，right 代表靠右对齐，top 代表靠上对齐；bottom 代表靠下对齐，center 代表居中对齐。

android:visible：视图的可视属性；visible 代表可见，invisible 代表不可见，占据视图布局位置；gone 代表不可见，不占据视图布局位置。

android:padding：内部内容与视图的距离。
- paddingTop 代表视图边缘与上边距离。
- paddingBottom 代表视图边缘与下边距离。
- paddingLeft 代表视图边缘与左边距离。
- paddingRight 代表视图边缘与右边距离。

app:layout_constraint：布局限制。
- app:layout_constraintBottom_toBottomOf 表示限制视图与底部对齐。
- app:layout_constraintTop_toTopOf 表示限制视图与顶部对齐。
- app:layout_constraintStart_ toStartOf 表示限制视图与左边对齐。
- app:layout_constraintEnd_toEndOf 限制视图与右边对齐。

View 的布局属性可以在代码中使用表 3-2 中的方法进行设置。

表 3-2 View 组件的属性设置方法列表

序号	方法名称	方法含义
1	setLayoutParams	设置视图的布局参数
2	setMinimumWidth	设置视图的最小宽度
3	setMinimumHeight	设置视图的最小高度
4	setBackgroundColor	设置视图的背景颜色
5	setBackgroundDrawable	设置视图的背景图片
6	setBackgroundResource	设置视图的背景资源
7	setPadding	设置视图边缘与视图内容之间的空白距离
8	setVisibility	设置视图的可视类型

Android 布局
（实践案例）

3.1.1 绝对布局

绝对布局也叫坐标布局，指定控件的绝对位置，简单直接，直观性强，但是手机屏幕尺寸差别较大，适应性差，Android 已弃用，可以用 Relative Layout（相对布局）替代。

如图 3-1 所示的是一个九宫格的布局，布局使用绝对布局来实现。视图中的每个组件使用的都是绝对布局，其中所有的组件高度和宽度都为绝对值，都是 80dp；组件 1 在空间中的位置也为绝对值，x 方向为 20dp，y 方向为 20dp；组件 2 在空间中的位置也采用绝对值，x 方向为 120dp，y 方向为 20dp；组件 3 在空间中的位置也采用绝对值，x 方向为 220dp，y 方向为 20dp；绝对布局的优点是布局简单，一目了然，但是缺点是布局不能适应屏幕的大小。

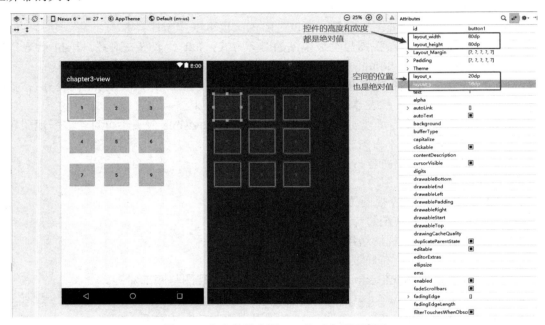

图 3-1　九宫格的布局——绝对布局示意图

3.1.2 相对布局

相对布局（Relative Layout）是 Android 用户界面设计使用得很普遍的布局类型。和其他布局很相似，相对布局可以通过 XML 布局资源来定义，也可以用 Java 程序来定义。相对布局的功能就像它的名字所表达的一样：它相对其他控件或父控件本身来组织控件。

其子控件，比如 ImageView、TextView 和 Button 控件，可以放在另外一个控件的上面、下面，或是左边、右边。子控件可以相对于父控件（相对布局容器）放置，包括放置在布局的顶部、底部、左部或右部边缘。在相对布局中，子控件的位置是相对兄弟控件或父容器而决定的。出于性能考虑，在设计相对布局时要按照控件之间的依赖关系排列，如 View A 的位置相对于 View B 来决定，则需要保证在布局文件中 View B 在 View A 的前面。

相对布局的常用属性及对应方法说明如表 3-3 所示。

表 3-3 Relative Layout 常用属性及对应方法说明

属性分类	属性	具体的说明描述
1. 属性值为 true 或 false	android:layout_centerHrizontal	水平居中
	android:layout_centerVertical	垂直居中
	android:layout_centerInparent	相对于父元素完全居中
	android:layout_alignParentBottom	贴紧父元素的下边缘
	android:layout_alignParentLeft	贴紧父元素的左边缘
	android:layout_alignParentRight	贴紧父元素的右边缘
	android:layout_alignParentTop	贴紧父元素的上边缘
	android:layout_alignWithParentIfMissing	对应的元素找不到的话就以父元素做参照物
2. 属性值必须为 id 的引用名 "@id/id-name"	android:layout_below	在某元素的下方
	android:layout_above	在某元素的的上方
	android:layout_toLeftOf	在某元素的左边
	android:layout_toRightOf	在某元素的右边
	android:layout_alignTop	本元素的上边缘和某元素的的上边缘对齐
	android:layout_alignLeft	本元素的左边缘和某元素的的左边缘对齐
	android:layout_alignBottom	本元素的下边缘和某元素的的下边缘对齐
	android:layout_alignRight	本元素的右边缘和某元素的的右边缘对齐
3. 属性值为具体的像素值，如 30dip，40px	android:layout_marginBottom	离某元素底边缘的距离
	android:layout_marginLeft	离某元素左边缘的距离
	android:layout_marginRight	离某元素右边缘的距离
	android:layout_marginTop	离某元素上边缘的距离

图 3-2 所示的是利用相对布局来实现九宫格的布局。布局使用相对布局来实现，视图中的每个组件都使用了相对布局，其中所有的组件高度和宽度值都为绝对值，都是 80dp。

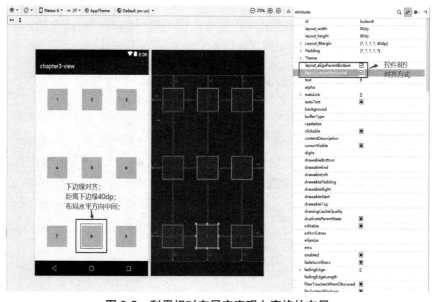

图 3-2 利用相对布局来实现九宫格的布局

组件 1、2、3 与顶部对齐并且与顶部的距离为 20dp；组件 4、5、6 在屏幕垂直方向上居于中间；组件 7、8、9 与底部对齐并且与底部距离为 20dp。

组件 1、4、7 与左边对齐并且与左边的距离为 20dp；组件 2、5、8 在屏幕水平方向上居于中间；组件 3、6、9 与右边对齐并且与右边的距离为 20dp。

组件之间的关系如图 3-3 所示。

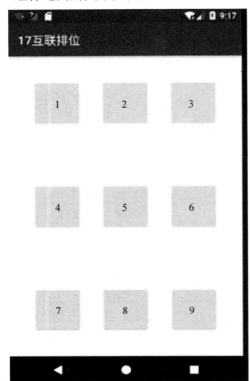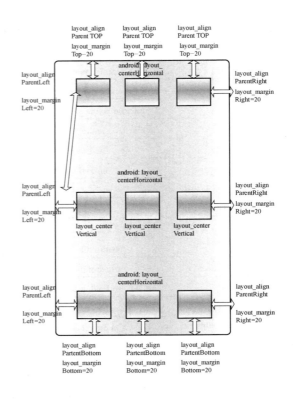

图 3-3　组件之间的关系

相对布局的优点是布局可以适应屏幕的位置；但是组件的大小是固定的，不能根据屏幕的大小动态改变。相对布局的代码如图 3-4 所示。

```xml
<?xml version="1.0" encoding="utf-8"?>
<RelativeLayout xmlns:android="http://schemas.android.com/apk/res/android"
    xmlns:app="http://schemas.android.com/apk/res-auto"
    android:layout_width="match_parent"
    android:layout_height="match_parent"
    android:layout_centerHorizontal="true">
    <Button
        android:id="@+id/button1"
        android:layout_width="80dp"
        android:layout_height="80dp"
        android:layout_alignParentStart="true"
        android:layout_alignParentLeft="true"
        android:layout_alignParentTop="true"
        android:layout_marginStart="40dp"
        android:layout_marginLeft="40dp"
        android:layout_marginTop="40dp"
        android:text="1" />
    <Button
        android:id="@+id/button2"
        android:layout_width="80dp"
```

图 3-4　相对布局的代码

```xml
        android:layout_height="80dp"
        android:layout_alignParentTop="true"
        android:layout_centerHorizontal="true"
        android:layout_marginTop="40dp"
        android:text="2" />
    <Button
        android:id="@+id/button3"
        android:layout_width="80dp"
        android:layout_height="80dp"
        android:layout_alignParentEnd="true"
        android:layout_alignParentRight="true"
        android:layout_alignParentTop="true"
        android:layout_marginRight="40dp"
        android:layout_marginTop="40dp"
        android:text="3" />
    <Button
        android:id="@+id/button4"
        android:layout_width="80dp"
        android:layout_height="80dp"
        android:layout_alignParentLeft="true"
        android:layout_alignParentStart="true"
        android:layout_centerVertical="true"
        android:layout_marginLeft="40dp"
        android:text="4" />
    <Button
        android:id="@+id/button5"
        android:layout_width="80dp"
        android:layout_height="80dp"
        android:layout_centerVertical="true"
        android:layout_centerHorizontal="true"
        android:text="5" />
    <Button
        android:id="@+id/button6"
        android:layout_width="80dp"
        android:layout_height="80dp"
        android:layout_alignParentEnd="true"
        android:layout_alignParentRight="true"
        android:layout_centerVertical="true"
        android:layout_marginRight="40dp"
        android:text="6" />
    <Button
        android:id="@+id/button7"
        android:layout_width="80dp"
        android:layout_height="80dp"
        android:layout_alignParentBottom="true"
        android:layout_alignParentLeft="true"
        android:layout_alignParentStart="true"
        android:layout_marginBottom="40dp"
        android:layout_marginLeft="40dp"
        android:text="7" />
    <Button
        android:id="@+id/button8"
        android:layout_width="80dp"
        android:layout_height="80dp"
        android:layout_alignParentBottom="true"
        android:layout_centerHorizontal="true"
        android:layout_marginBottom="40dp"
        android:text="8" />
    <Button
        android:id="@+id/button9"
        android:layout_width="80dp"
        android:layout_height="80dp"
        android:layout_alignParentBottom="true"
        android:layout_alignParentEnd="true"
        android:layout_alignParentRight="true"
        android:layout_marginBottom="40dp"
        android:layout_marginRight="40dp"
        android:text="9" />
</RelativeLayout>
```

图 3-4 相对布局的代码（续）

3.1.3 线性布局

线性布局（Linear Layout），从外框上可以理解为一个 div（分隔符），它是一个一个地从上往下罗列在屏幕上的。每一个 Linear Layout 里面又可分为垂直布局（android:orientation="vertical"）和水平布局（android:orientation="horizontal"）。当垂直布局时，每一行就只有一个元素，多个元素依次垂直往下；水平布局时，只有一行，每一个元素依次向右排列。

Linear Layout 按照垂直或者水平的顺序依次排列子元素，每一个子元素都位于前一个元素之后。如果是垂直排列的，那么将是一个 N 行单列的结构，每一行只会有一个元素，而不论这个元素的宽度为多少；如果是水平排列的，那么将是一个单行 N 列的结构。如果搭建两行两列的结构，通常的方式是先垂直排列两个元素，每一个元素里再包含一个 Linear Layout 进行水平排列。

LinearLayout 布局的属性既可以在布局文件（XML）中设置，也可以通过成员方法进行设置。表 3-4 给出了 Linear Layout 常用属性及对应方法说明。

表 3-4 Linear Layout 常用属性及对应方法说明

属性名	对应方法	具体的描述
android:orientation	setOrientation(int)	设置线性布局的朝向，可取 horizontal 和 vertical 两种排列方式
android:gravity	setGravity(int)	设置线性布局的内部元素的布局方式

Linear Layout 中的子元素属性 android:layout_weight 生效，它用于描述该子元素在剩余空间中占有的大小比例。加入一行只有一个文本框，那么它的默认值就为 0，如果一行中有两个等长的文本框，那么它们的 android:layout_weight 值可以同为 1。如果一行中有两个不等长的文本框，那么它们的 android:layout_weight 值分别为 1 和 2，即第一个文本框将占据剩余空间的三分之二，第二个文本框将占据剩余空间中的三分之一。android:layout_weight 遵循"数值越小，重要度越高"的原则。

图 3-5 所示的是使用线性布局来实现的九宫格布局；视图中的每个组件都是使用线性布局来实现的，其中所有的组件高度和宽度都是根据屏幕的大小来变化的；线性布局使子视图像线一样串起来；布局属性中"android:orientation="horizontal""代表从左到右排列；"android:orientation= "vertical""代表从上到下排列。

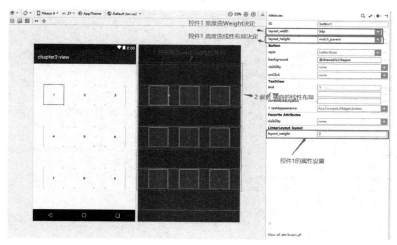

图 3-5 使用线性布局来实现的九宫格布局

线性布局组件之间的关系如图 3-6 所示；组件 1、2、3 组成线性子布局 1；组件 4、5、6 组成线性子布局 2；组件 7、8、9 组成线性子布局 3；它们在垂直方向上的高度权重都是 2；几个线性子布局的间隔添加了空白 View，空白 View 的权重设置为 1；线性子布局 1 中，组件 1、2、3 从左到右排列；组件在水平方向上的高度权重都是 2；组件的间隔添加了空白 View，空白 View 的权重设置为 1。

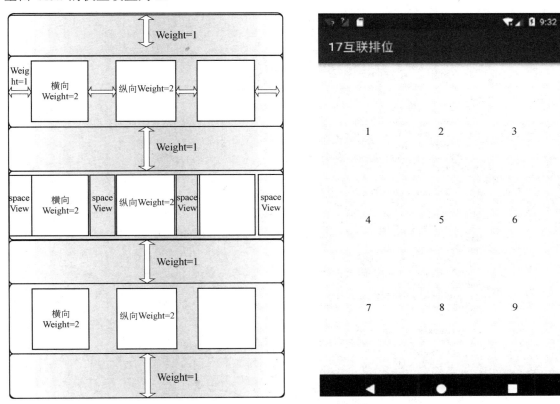

图 3-6　线性布局组件之间的关系

线性布局的代码如图 3-7 所示。

```xml
<?xml version="1.0" encoding="utf-8"?>
<LinearLayout xmlns:android="http://schemas.android.com/apk/res/android"
    android:layout_width="match_parent"
    android:layout_height="match_parent"
    android:orientation="vertical">
    <LinearLayout
        android:layout_width="match_parent"
        android:layout_height="0dp"
        android:layout_weight="1"
        android:orientation="horizontal"
        android:visibility="invisible">
    </LinearLayout>
    <LinearLayout
        android:layout_width="match_parent"
        android:layout_height="0dp"
        android:layout_weight="1"
        android:orientation="horizontal">
        <View
```

图 3-7　线性布局的代码

```xml
        android:layout_width="0dp"
        android:layout_height="match_parent"
        android:layout_weight="1"
        android:visibility="invisible" />
    <Button
        android:id="@+id/button1"
        android:layout_width="0dp"
        android:layout_height="match_parent"
        android:layout_weight="2"
        android:background="@drawable/shapes"
        android:text="1" />
    <View
        android:layout_width="0dp"
        android:layout_height="match_parent"
        android:layout_weight="1"
        android:visibility="invisible" />
    <Button
        android:id="@+id/button2"
        android:layout_width="0dp"
        android:layout_height="match_parent"
        android:layout_weight="2"
        android:background="@drawable/shapes"
        android:text="2" />
    <View
        android:layout_width="0dp"
        android:layout_height="match_parent"
        android:layout_weight="1"
        android:visibility="invisible" />
    <Button
        android:id="@+id/button3"
        android:layout_width="0dp"
        android:layout_height="match_parent"
        android:layout_weight="2"
        android:background="@drawable/shapes"
        android:text="3" />
    <View
        android:layout_width="0dp"
        android:layout_height="match_parent"
        android:layout_weight="1"
        android:visibility="invisible" />
</LinearLayout>
<LinearLayout
    android:layout_width="match_parent"
    android:layout_height="0dp"
    android:layout_weight="1"
    android:orientation="horizontal"
    android:visibility="invisible">
</LinearLayout>
<LinearLayout
    android:layout_width="match_parent"
    android:layout_height="0dp"
    android:layout_weight="1"
    android:orientation="horizontal">
    <View
        android:layout_width="0dp"
        android:layout_height="match_parent"
        android:layout_weight="1"
        android:visibility="invisible" />
    <Button
        android:layout_width="0dp"
        android:id="@+id/button4"
        android:layout_height="match_parent"
        android:layout_weight="2"
        android:background="@drawable/shapes"
        android:text="4" />
    <View
```

图 3-7 线性布局的代码（续）

```xml
            android:layout_width="0dp"
            android:layout_height="match_parent"
            android:layout_weight="1"
            android:visibility="invisible" />
        <Button
            android:layout_width="0dp"
            android:id="@+id/button5"
            android:layout_height="match_parent"
            android:layout_weight="2"
            android:background="@drawable/shapes"
            android:text="5" />
        <View
            android:layout_width="0dp"
            android:layout_height="match_parent"
            android:layout_weight="1"
            android:visibility="invisible" />
        <Button
            android:layout_width="0dp"
            android:id="@+id/button6"
            android:layout_height="match_parent"
            android:layout_weight="2"
            android:background="@drawable/shapes"
            android:text="6" />
        <View
            android:layout_width="0dp"
            android:layout_height="match_parent"
            android:layout_weight="1"
            android:visibility="invisible" />
    </LinearLayout>
    <LinearLayout
        android:layout_height="0dp"
        android:layout_width="match_parent"
        android:layout_weight="1"
        android:orientation="horizontal">
    </LinearLayout>
    <LinearLayout
        android:layout_width="match_parent"
        android:layout_height="0dp"
        android:layout_weight="1"
        android:orientation="horizontal">
        <View
            android:layout_width="0dp"
            android:layout_height="match_parent"
            android:layout_weight="1"
            android:visibility="invisible" />
        <Button
            android:layout_width="0dp"
            android:id="@+id/button7"
            android:layout_height="match_parent"
            android:layout_weight="2"
            android:background="@drawable/shapes"
            android:text="7" />
        <View
            android:layout_width="0dp"
            android:layout_height="match_parent"
            android:layout_weight="1"
            android:visibility="invisible" />
        <Button
            android:layout_width="0dp"
            android:id="@+id/button8"
            android:layout_height="match_parent"
            android:layout_weight="2"
            android:background="@drawable/shapes"
            android:text="8" />
        <View
            android:layout_width="0dp"
```

图 3-7 线性布局的代码（续）

```xml
            android:layout_height="match_parent"
            android:layout_weight="1"
            android:visibility="invisible" />
        <Button
            android:layout_width="0dp"
            android:id="@+id/button9"
            android:layout_height="match_parent"
            android:layout_weight="2"
            android:background="@drawable/shapes"
            android:text="9" />
        <View
            android:layout_width="0dp"
            android:layout_height="match_parent"
            android:layout_weight="1"
            android:visibility="invisible" />
    </LinearLayout>
    <LinearLayout
        android:layout_width="match_parent"
        android:layout_height="0dp"
        android:layout_weight="1"
        android:orientation="horizontal"
        android:visibility="invisible">

    </LinearLayout>
</LinearLayout>
```

图 3-7　线性布局的代码（续）

3.1.4　约束布局

从上面的线性布局中我们可以看到，实现复杂的布局十分困难且复杂，需要多层嵌套，UI 设计有时毫无规律且复杂；有时设计师的标注不够科学，往往只有简单标注元素之间的距离，缺少对齐方式和布局空间的概念，因此需要自己推敲实现，效率难以保证。

2016 年 Google 在 I/O 大会上推出了全新的布局控件 Constraint Layout 并得到官方的大力推广。Constraint Layout 中文直译就是约束布局，顾名思义就是通过对控件进行约束布局，简单理解就可以认为它是 Relative Layout 的超级加强升级版本。可以有效解决布局嵌套的性能问题和复杂布局的快速实现，性能大约提高 40%；Constraint Layout 实现布局的方法也十分简单，可以通过 Android Studio 提供的编辑器，简单拖曳就可以快速实现布局，就好像 UI 设计师操作 Photoshop 一样，熟练掌握以后，安装 UI 标注就可以直接还原设计效果。

约束布局定义了 13 个基本约束属性用来控制子 View 的相对位置，Constraint Layout 最基本的属性控制有 layout_constraintXXX_toYYYOf 格式的属性，即将 "View A" 的方向 XXX 置于 "View B" 的方向 YYY。当中，View B 可以是父容器即 Constraint Layout，用 "parent" 来表示。下面是各个布局属性的含义。

- ◆ layout_constraintBaseline_toBaselineOf：View A 内部文字与 View B 内部文字对齐。
- ◆ layout_constraintLeft_toLeftOf：View A 与 View B 左对齐。
- ◆ layout_constraintLeft_toRightOf：View A 的左边置于 View B 的右边。
- ◆ layout_constraintRight_toLeftOf：View A 的右边置于 View B 的左边。
- ◆ layout_constraintRight_toRightOf：View A 与 View B 右对齐。
- ◆ layout_constraintTop_toTopOf：View A 与 View B 顶部对齐。
- ◆ layout_constraintTop_toBottomOf：View A 的顶部置于 View B 的底部。
- ◆ layout_constraintBottom_toTopOf：View A 的底部置于 View B 的顶部。

- layout_constraintBottom_toBottomOf：View A 与 View B 底部对齐。
- layout_constraintStart_toEndOf：View A 的左边置于 View B 的右边。
- layout_constraintStart_toStartOf：View A 与 View B 左对齐。
- layout_constraintEnd_toStartOf：View A 的右边置于 View B 的左边。
- layout_constraintEnd_toEndOf：View A 与 View B 右对齐。

图 3-8 所示的是 Constraint Layout 布局的一个实例，在约束布局中定义了一个 View，View 与父容器（整个布局）的约束关系如下。

- layout_constraintStart_toStartOf：View 与父容器左对齐。
- layout_constraintEnd_toEndOf：View 与父容器右对齐。
- layout_constraintTop_toTopOf：View 与父容器顶部对齐。
- layout_constraintBottom_toBottomOf：View 与父容器底部对齐。

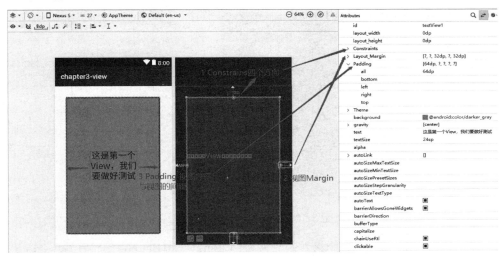

图 3-8 约束布局的一个案例

View 与父容器之间有边距，View 与父容器的前、后、左、右的边距均设置为 32dp；View 中的文字在 View 视图中居中,文字与视图的距离设置为 64dp；约束布局的代码如图 3-9 所示。

```xml
<?xml version="1.0" encoding="utf-8"?>
<android.support.constraint.ConstraintLayout
    xmlns:android="http://schemas.android.com/apk/res/android"
    xmlns:app="http://schemas.android.com/apk/res-auto"
    xmlns:tools="http://schemas.android.com/tools"
    android:layout_width="match_parent"
    android:layout_height="match_parent">
    <TextView
        android:id="@+id/textView1"
        android:layout_width="0dp"
        android:layout_height="0dp"
        android:layout_marginStart="32dp"
        android:layout_marginTop="32dp"
        android:layout_marginEnd="32dp"
        android:layout_marginBottom="32dp"
        android:gravity="center"
        android:padding="64dp"
        android:text="这是第一个 View，我们要做好测试"
        android:textSize="24sp"
```

图 3-9 约束布局的代码

```
        app:layout_constraintBottom_toBottomOf="parent"
        app:layout_constraintEnd_toEndOf="parent"
        app:layout_constraintHorizontal_bias="0.0"
        app:layout_constraintStart_toStartOf="parent"
        app:layout_constraintTop_toTopOf="parent"
        app:layout_constraintVertical_bias="0.0" />
</android.support.constraint.ConstraintLayout>
```

图 3-9 约束布局的代码（续）

在上一个实例中，如果减少一个约束条件，比如 View 组件在 Constraint Layout 布局的约束条件只有上、左、右三个方向，那么 View 组件的高度可以设置为由内容决定，如图 3-10 所示。

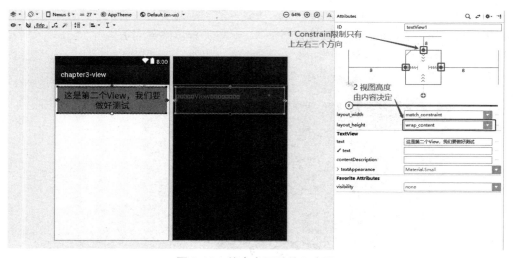

图 3-10 约束布局组件示意图

图 3-11 所示的是一个具有多个组件的约束实例，Button1 和 Button2 组件在 Constraint Layout 布局的约束条件只有上、左、右三个方向，高度设置为内容适配，也就是 Button 中的文字有多少就占据多大的高度；Button3 组件在 Constraint Layout 布局的约束条件有上、下、左、右 4 个方向，高度设置为 0；那么 View 组件的高度设置由约束来决定，高度由屏幕的剩余的大小决定。

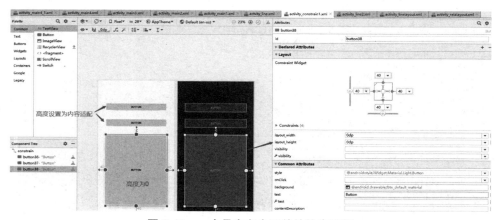

图 3-11 一个具有多个组件的约束实例

Android 中的线性布局可以通过各种 Weight 比重来实现，但会导致 Linear Layout 各种嵌套层级越来越深，不易维护又影响性能。后来还有过百分比布局，但是相比而言 Andorid 的 Guideline 高效又实用。如图 3-12 所示的是 Guideline 的应用实例。

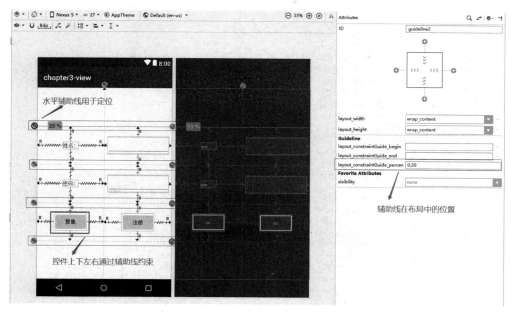

图 3-12　Guideline 的应用实例

拖了数个横竖的 Guideline 后，界面的整体框架就会出来了，再根据需要把相应的控件约束到对应 Guideline。UI 设计师可以把整体的布局直接切图，不用再把每个小布局都切出来，程序员可以根据内容拉出几根横竖的 Guideline；由于 Guideline 是根据百分比平分界面的，控件又约束于这些 Guideline，也就完成了适配。Guideline 的样例代码如图 3-13 所示。

```xml
<?xml version="1.0" encoding="utf-8"?>
<android.support.constraint.ConstraintLayout xmlns:android="http://schemas.android.com/apk/res/android"
    xmlns:app="http://schemas.android.com/apk/res-auto"
    xmlns:tools="http://schemas.android.com/tools"
    android:layout_width="match_parent"
    android:layout_height="match_parent">
    <android.support.constraint.Guideline
        android:id="@+id/guideline"
        android:layout_width="wrap_content"
        android:layout_height="wrap_content"
        android:orientation="vertical"
        app:layout_constraintGuide_percent="0.5"
        app:layout_constraintTop_toTopOf="parent" />
    <Button
        android:id="@+id/button"
        android:layout_width="0dp"
        android:layout_height="wrap_content"
        android:layout_marginBottom="24dp"
        android:layout_marginEnd="24dp"
        android:layout_marginLeft="24dp"
        android:layout_marginRight="24dp"
        android:layout_marginStart="24dp"
        android:layout_marginTop="24dp"
        android:text="登录"
```

图 3-13　Guideline 的样例代码

```xml
        app:layout_constraintBottom_toTopOf="@+id/guideline5"
        app:layout_constraintDimensionRatio="3:1"
        app:layout_constraintEnd_toStartOf="@+id/guideline"
        app:layout_constraintStart_toStartOf="parent"
        app:layout_constraintTop_toTopOf="@+id/guideline4" />
<Button
    android:id="@+id/button3"
    android:layout_width="wrap_content"
    android:layout_height="wrap_content"
    android:layout_marginBottom="8dp"
    android:layout_marginEnd="8dp"
    android:layout_marginLeft="8dp"
    android:layout_marginRight="8dp"
    android:layout_marginStart="8dp"
    android:layout_marginTop="8dp"
    android:text="注册"
    app:layout_constraintBottom_toTopOf="@+id/guideline5"
    app:layout_constraintEnd_toEndOf="parent"
    app:layout_constraintStart_toStartOf="@+id/guideline"
    app:layout_constraintTop_toTopOf="@+id/guideline4" />
<android.support.constraint.Guideline
    android:id="@+id/guideline2"
    android:layout_width="wrap_content"
    android:layout_height="wrap_content"
    android:orientation="horizontal"
    app:layout_constraintGuide_percent="0.20" />
<android.support.constraint.Guideline
    android:id="@+id/guideline3"
    android:layout_width="wrap_content"
    android:layout_height="wrap_content"
    android:orientation="horizontal"
    app:layout_constraintGuide_percent="0.4" />
<android.support.constraint.Guideline
    android:id="@+id/guideline4"
    android:layout_width="wrap_content"
    android:layout_height="wrap_content"
    android:orientation="horizontal"
    app:layout_constraintGuide_percent="0.6" />
<android.support.constraint.Guideline
    android:id="@+id/guideline5"
    android:layout_width="wrap_content"
    android:layout_height="wrap_content"
    android:orientation="horizontal"
    app:layout_constraintGuide_percent="0.8" />
<TextView
    android:id="@+id/textView"
    android:layout_width="wrap_content"
    android:layout_height="wrap_content"
    android:layout_marginStart="8dp"
    android:layout_marginLeft="8dp"
    android:layout_marginTop="8dp"
    android:layout_marginEnd="8dp"
    android:layout_marginRight="8dp"
    android:layout_marginBottom="8dp"
    android:text="姓名+学号"
    android:textSize="24sp"
    app:layout_constraintBottom_toTopOf="@+id/guideline3"
    app:layout_constraintEnd_toStartOf="@+id/guideline"
    app:layout_constraintStart_toStartOf="parent"
    app:layout_constraintTop_toTopOf="@+id/guideline2" />
<TextView
    android:id="@+id/textView2"
    android:layout_width="wrap_content"
    android:layout_height="wrap_content"
    android:layout_marginBottom="8dp"
    android:layout_marginEnd="8dp"
```

图 3-13　Guideline 的样例代码（续）

```xml
        android:layout_marginLeft="8dp"
        android:layout_marginRight="8dp"
        android:layout_marginStart="8dp"
        android:layout_marginTop="8dp"
        android:text="密码："
        app:layout_constraintBottom_toTopOf="@+id/guideline4"
        app:layout_constraintEnd_toStartOf="@+id/guideline"
        app:layout_constraintStart_toStartOf="parent"
        app:layout_constraintTop_toTopOf="@+id/guideline3" />
    <EditText
        android:id="@+id/editText"
        android:layout_width="0dp"
        android:layout_height="wrap_content"
        android:layout_marginBottom="8dp"
        android:layout_marginEnd="8dp"
        android:layout_marginLeft="8dp"
        android:layout_marginRight="8dp"
        android:layout_marginStart="8dp"
        android:layout_marginTop="8dp"
        android:ems="10"
        android:inputType="textPersonName"
        app:layout_constraintBottom_toTopOf="@+id/guideline3"
        app:layout_constraintEnd_toEndOf="parent"
        app:layout_constraintStart_toStartOf="@+id/guideline"
        app:layout_constraintTop_toTopOf="@+id/guideline2" />
    <EditText
        android:id="@+id/editText2"
        android:layout_width="0dp"
        android:layout_height="wrap_content"
        android:layout_marginBottom="8dp"
        android:layout_marginEnd="8dp"
        android:layout_marginLeft="8dp"
        android:layout_marginRight="8dp"
        android:layout_marginStart="8dp"
        android:layout_marginTop="8dp"
        android:ems="10"
        android:inputType="textPassword"
        app:layout_constraintBottom_toTopOf="@+id/guideline4"
        app:layout_constraintEnd_toEndOf="parent"
        app:layout_constraintStart_toStartOf="@+id/guideline"
        app:layout_constraintTop_toTopOf="@+id/guideline3" />
</android.support.constraint.ConstraintLayout>
```

图 3-13　Guideline 的样例代码（续）

根据 Guideline 的布局，可以自己设计一个约束布局，如图 3-14 所示。

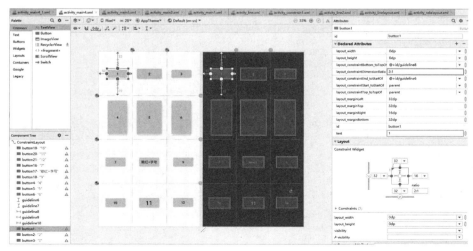

图 3-14　根据 Guideline 的布局设计的约束布局

约束布局支持组件之间通过 Chains 来进行布局，Chains 简单来说就是用一条链将同一轴（水平或者垂直）的各个组件连接起来，使它们能够统一行动。如图 3-15 所示的是一个利用 Chains 进行布局的实例。

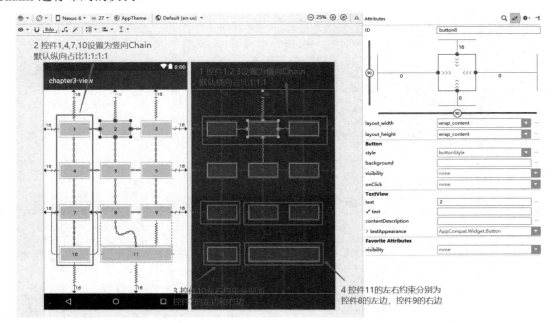

图 3-15　一个利用 Chains 进行布局的实例

在 Chains 组件布局中，我们可以将组件 1、2、3 分为一组，4、5、6 分为一组；7、8、9 分为一组，设置为横向的 Chains。组件 1、4、7、10 分为一组，2、5、8、11 分为一组；3、6、9 分为一组，设置为纵向的 Chains。Chains 默认的样式为 spread，组件被分开，并且组件的比例为 1∶1∶1；对于组件 10 来说，左边与组件 7 的左边对齐，右边与组件 7 的右边对齐；对于组件 11 来说，左边与组件 8 的左边对齐，右边与组件 9 的右边对齐；Chains 组件布局的代码如图 3-16 所示。

```xml
<?xml version="1.0" encoding="utf-8"?>
<android.support.constraint.ConstraintLayout xmlns:android="http://schemas.android.com/apk/res/android"
    xmlns:app="http://schemas.android.com/apk/res-auto"
    xmlns:tools="http://schemas.android.com/tools"
    android:layout_width="match_parent"
    android:layout_height="match_parent">

    <Button
        android:id="@+id/button1"
        android:layout_width="wrap_content"
        android:layout_height="wrap_content"
        android:layout_marginStart="16dp"
        android:layout_marginLeft="16dp"
        android:layout_marginTop="16dp"
        android:text="1"
        app:layout_constraintBottom_toTopOf="@+id/button4"
        app:layout_constraintEnd_toStartOf="@+id/button2"
        app:layout_constraintHorizontal_bias="1"
        app:layout_constraintStart_toStartOf="parent"
        app:layout_constraintTop_toTopOf="parent" />
```

图 3-16　Chains 组件布局的代码

```xml
<Button
    android:id="@+id/button2"
    android:layout_width="wrap_content"
    android:layout_height="wrap_content"
    android:layout_marginTop="16dp"
    android:text="2"
    app:layout_constraintBottom_toTopOf="@+id/button5"
    app:layout_constraintEnd_toStartOf="@+id/button3"
    app:layout_constraintHorizontal_bias="2"
    app:layout_constraintStart_toEndOf="@+id/button1"
    app:layout_constraintTop_toTopOf="parent" />

<Button
    android:id="@+id/button3"
    android:layout_width="wrap_content"
    android:layout_height="wrap_content"
    android:layout_marginTop="16dp"
    android:layout_marginEnd="16dp"
    android:layout_marginRight="16dp"
    android:text="3"
    app:layout_constraintBottom_toTopOf="@+id/button6"
    app:layout_constraintEnd_toEndOf="parent"
    app:layout_constraintHorizontal_bias="1"
    app:layout_constraintStart_toEndOf="@+id/button2"
    app:layout_constraintTop_toTopOf="parent" />

<Button
    android:id="@+id/button4"
    android:layout_width="wrap_content"
    android:layout_height="wrap_content"
    android:layout_marginLeft="16dp"
    android:layout_marginStart="16dp"
    android:text="4"
    app:layout_constraintBottom_toTopOf="@+id/button7"
    app:layout_constraintEnd_toStartOf="@+id/button5"
    app:layout_constraintHorizontal_bias="0.5"
    app:layout_constraintStart_toStartOf="parent"
    app:layout_constraintTop_toBottomOf="@+id/button1" />

<Button
    android:id="@+id/button5"
    android:layout_width="wrap_content"
    android:layout_height="wrap_content"
    android:text="5"
    app:layout_constraintBottom_toTopOf="@+id/button8"
    app:layout_constraintEnd_toStartOf="@+id/button6"
    app:layout_constraintHorizontal_bias="0.5"
    app:layout_constraintStart_toEndOf="@+id/button4"
    app:layout_constraintTop_toBottomOf="@+id/button2" />

<Button
    android:id="@+id/button6"
    android:layout_width="wrap_content"
    android:layout_height="wrap_content"
    android:layout_marginEnd="16dp"
    android:layout_marginRight="16dp"
    android:text="6"
    app:layout_constraintBottom_toTopOf="@+id/button9"
    app:layout_constraintEnd_toEndOf="parent"
    app:layout_constraintHorizontal_bias="0.5"
    app:layout_constraintStart_toEndOf="@+id/button5"
    app:layout_constraintTop_toBottomOf="@+id/button3" />

<Button
```

图 3-16　Chains 组件布局的代码（续）

```xml
    android:id="@+id/button7"
    android:layout_width="wrap_content"
    android:layout_height="wrap_content"
    android:layout_marginLeft="16dp"
    android:layout_marginStart="16dp"
    android:text="7"
    app:layout_constraintBottom_toTopOf="@+id/button10"
    app:layout_constraintEnd_toStartOf="@+id/button8"
    app:layout_constraintHorizontal_bias="0.5"
    app:layout_constraintStart_toStartOf="parent"
    app:layout_constraintTop_toBottomOf="@+id/button4" />

<Button
    android:id="@+id/button8"
    android:layout_width="wrap_content"
    android:layout_height="wrap_content"
    android:text="8"
    app:layout_constraintBottom_toTopOf="@+id/button11"
    app:layout_constraintEnd_toStartOf="@+id/button9"
    app:layout_constraintHorizontal_bias="0.5"
    app:layout_constraintStart_toEndOf="@+id/button7"
    app:layout_constraintTop_toBottomOf="@+id/button5" />

<Button
    android:id="@+id/button10"
    android:layout_width="wrap_content"
    android:layout_height="wrap_content"
    android:layout_marginBottom="16dp"
    android:text="10"
    app:layout_constraintBottom_toBottomOf="parent"
    app:layout_constraintEnd_toEndOf="@+id/button7"
    app:layout_constraintHorizontal_bias="0.5"
    app:layout_constraintStart_toStartOf="@+id/button7"
    app:layout_constraintTop_toBottomOf="@+id/button7" />

<Button
    android:id="@+id/button11"
    android:layout_width="0dp"
    android:layout_height="wrap_content"
    android:layout_marginBottom="16dp"
    android:text="赖红+学号"
    app:layout_constraintBottom_toBottomOf="parent"
    app:layout_constraintEnd_toEndOf="@+id/button9"
    app:layout_constraintHorizontal_bias="4.0"
    app:layout_constraintStart_toStartOf="@+id/button8"
    app:layout_constraintTop_toBottomOf="@+id/button8" />

<Button
    android:id="@+id/button9"
    android:layout_width="wrap_content"
    android:layout_height="wrap_content"
    android:layout_marginEnd="16dp"
    android:layout_marginRight="16dp"
    android:text="9"
    app:layout_constraintBottom_toTopOf="@+id/button11"
    app:layout_constraintEnd_toEndOf="parent"
    app:layout_constraintHorizontal_bias="0.5"
    app:layout_constraintStart_toEndOf="@+id/button8"
    app:layout_constraintTop_toBottomOf="@+id/button6" />
</android.support.constraint.ConstraintLayout>
```

图 3-16 Chains 组件布局的代码（续）

Chains 也支持权重的设置，在上一个实例中，可以设置组件的高度为 0，即 android:layout_height="0dp"，然后再设置组件的权重，即 app:layout_constraintVertical_weight="1"。如图 3-17

所示的是一个 Chains 中设置权重的实例。在 Chains 中，1、4、7、10 四个组件，分别设置组件 1 和 4 的高度权重为 1，组件 7 和 10 的高度权重为 0.5。可以看出，组件 1 和 4 的高度是组件 7 和 10 的两倍。

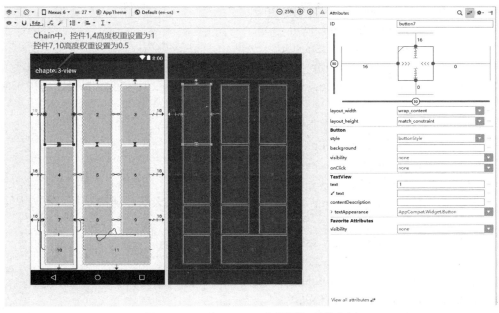

图 3-17　一个 Chains 中设置权重的实例

3.1.5　单元小测

选择题：

1. 下面哪一个属性是与父视图一样的？（　　）
A. match_parent　　　B. wrap_content　　　C. match_content　　　D. wrap_parent

2. 下面哪一个属性是根据内容调整大小的？（　　）
A. match_parent　　　B. wrap_content　　　C. match_content　　　D. wrap_parent

3. 下面哪一个属性用于表示视图与周围视图的距离？（　　）
A. android:layout_width　　　　　B. android:layout_height
C. android:layout_margin　　　　D. android:padding

4. 下面哪一个属性用于表示视图与上级视图的对齐方式？（　　）
A. android:layout_gravity　　　　B. android:gravity

5. 下面哪一个属性用于表示视图中文字与视图的对齐方式？（　　）
A. android:layout_gravity　　　　B. android:gravity

6. 视图的可视属性哪一个表示消失，不占用位置？（　　）
A. visible　　　　　B. invisible　　　　　C. gone

7. 视图的可视属性哪一个表示消失，占用位置？（　　）
A. visible　　　　　B. invisible　　　　　C. gone

8. 下面哪一个属性用于表示视图内部内容与视图的距离？（　　）

A. android:layout_width　　　　　　B. android:layout_height
C. android:layout_margin　　　　　　D. android:padding

9. 在限制性布局中，app:layout_constraint 属性设置为什么值才能限制视图与右边对齐？（　　）

A. app:layout_constraintBottom_toBottomOf

B. app:layout_constraintEnd_toEndOf

C. app:layout_constraintStart_toStartOf

D. app:app：layout_constraintTop_toTopOf

10. 在限制性布局中，app:layout_constraint 属性设置为什么值才能限制视图与顶部对齐？（　　）

A. app:layout_constraintBottom_toBottomOf

B. app:layout_constraintEnd_toEndOf

C. app:layout_constraintStart_toStartOf

D. app:layout_constraintTop_toTopOf

11. 在限制性布局中，app:layout_constraint 属性设置为什么值才能限制视图与左边对齐？（　　）

A. app:layout_constraintBottom_toBottomOf

B. app:layout_constraintEnd_toEndOf

C. app:layout_constraintStart_toStartOf

D. app:layout_constraintTop_toTopOf

12. 在限制性布局中，app:layout_constraint 属性设置为什么值才能限制视图与底部对齐？（　　）

A. app:layout_constraintBottom_toBottomOf

B. app:layout_constraintEnd_toEndOf

C. app:layout_constraintStart_toStartOf

D. app:layout_constraintTop_toTopOf

13. Android 的所有控件都继承自哪个类？（　　）

A. Control　　　B. Window　　　C. Activity　　　D. View

14. Android 的所有布局都继承自哪个类？（　　）

A. Layout　　　B. ViewGroup　　　C. Container　　　D. View

3.2　Android 基础组件

基础组件（慕课）

本节介绍 Android 最基础的几个组件的使用方法，主要包括文本视图 TextView 组件、按钮 Button 组件、编辑框组件 EditText，进度条组件 ProgressBar。

3.2.1　TextView

文本视图 TextView 组件是最基础的文本显示组件。其 XML 属性列表和设置方法如表 3-5 所示。

表 3-5　TextView 组件的 XML 属性列表和设置方法

序号	XML 属性	TextView 类设置方法	说明
1	android:text	setText	设置文本内容
2	android:textColor	setTextColor	设置文本颜色
3	android:textSize	setTextSize	设置文本大小
4	android:textAppearance	setTextAppearance	设置文本样式 res/styles.xml
5	android:gravity	setGravity	设置文本的对齐方式
6	android:singleLine	setSingleLine	设置文本的单行显示
7	android:ellipsize	setEllipsize	设置文本超出范围的省略方式
8	android:focusable	setFocusable	设置是否获得焦点

android:ellipsize 这个属性用于设置文本超出范围的省略方式，省略方式的值如表 3-6 所示。

表 3-6　android:ellipsize 属性省略方式的值

序号	XML 属性值	方法设置值	说明
1	start	START	省略号在开头
2	middle	MIDDLE	省略号在中间
3	end	END	省略号在末尾
4	marquee	MARQUEE	跑马灯显示

如图 3-18 所示的是使用 TextView 完成天气预报的动态显示的一个具体实例，由于手机屏幕有限，一般的新闻内容比较长，TextView 不能完全显示。因此，可以设置 TextView 为跑马灯显示的方式。

图 3-18　使用 TextView 完成天气预报的动态显示

TextView 的 XML 布局关键的属性设置如图 3-19 所示。

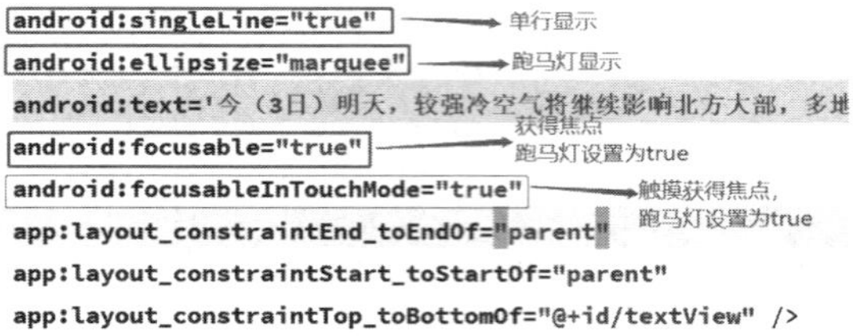

```
android:singleLine="true"        → 单行显示
android:ellipsize="marquee"      → 跑马灯显示
android:text='今（3日）明天，较强冷空气将继续影响北方大部，多地
android:focusable="true"         → 跑马灯设置为true
android:focusableInTouchMode="true"  → 触摸获得焦点，跑马灯设置为true
app:layout_constraintEnd_toEndOf="parent"
app:layout_constraintStart_toStartOf="parent"
app:layout_constraintTop_toBottomOf="@+id/textView" />
```

图 3-19　TextView 的 XML 布局关键属性设置

android:text="天气信息"用于设置文本内容；android:singleLine="true"用于设置文本的单行显示。

android:ellipsize="marquee"用于设置文本以跑马灯方式显示；android:focusable="true"用于设置跑马灯获得焦点；android:focusableInTouchMode="true"用于设置触摸获得焦点。以跑马灯方式显示天气预报的代码如图 3-20 所示。

```xml
<?xml version="1.0" encoding="utf-8"?>
<android.support.constraint.ConstraintLayout
    xmlns:android="http://schemas.android.com/apk/res/android"
    xmlns:app="http://schemas.android.com/apk/res-auto"
    xmlns:tools="http://schemas.android.com/tools"
    android:layout_width="match_parent"
    android:layout_height="match_parent">

    <TextView
        android:id="@+id/textView"
        android:layout_width="wrap_content"
        android:layout_height="wrap_content"
        android:layout_marginEnd="16dp"
        android:layout_marginLeft="16dp"
        android:layout_marginRight="16dp"
        android:layout_marginStart="16dp"
        android:layout_marginTop="16dp"
        android:text="跑马灯实例"
        app:layout_constraintEnd_toEndOf="parent"
        app:layout_constraintStart_toStartOf="parent"
        app:layout_constraintTop_toTopOf="parent" />

    <TextView
        android:id="@+id/txt_news"
        android:layout_width="0dp"
        android:layout_height="wrap_content"
        android:layout_marginStart="16dp"
        android:layout_marginLeft="16dp"
        android:layout_marginTop="16dp"
        android:layout_marginEnd="16dp"
        android:layout_marginRight="16dp"
        android:ellipsize="marquee"
        android:focusable="true"
        android:focusableInTouchMode="true"
        android:singleLine="true"
        android:text='今（3日）明天，较强冷空气将继续影响北方大部，多地气温下降 6℃左右，西北地区、华北、东北将现大范围雨雪。同时，冷空气也将清除京津冀一带的雾和霾。随着台风"玉兔"减弱消散。'
        android:textSize="20dp"
        app:layout_constraintEnd_toEndOf="parent"
        app:layout_constraintStart_toStartOf="parent"
        app:layout_constraintTop_toBottomOf="@+id/textView" />
</android.support.constraint.ConstraintLayout>
```

图 3-20　以跑马灯方式显示天气预报的代码

3.2.2 Button

按钮（Button）组件是最常用的 UI 交互组件，是 TextView 的派生类，区别是 Button 有一个比较明显的按钮外观，用户可以单击，如果去除 Button 的 XML 属性 backgroud 为@null，则两者外观基本一致。如图 3-21 所示的是 Button 的一个实例。单击按钮，提示按钮被单击；长按按钮，提示按钮被长按了。

图 3-21　Button 的一个实例

用户单击 Button 后会产生事件，Android 系统如何监听到事件并处理呢？Android 提供了一整套的组件事件响应机制来处理不同的组件产生的事件，如图 3-22 所示的是 Android 的事件响应原理。

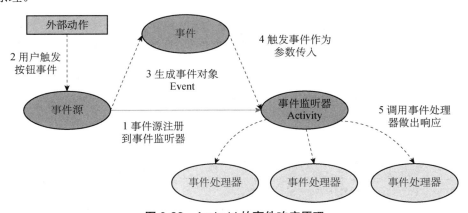

图 3-22　Android 的事件响应原理

事件处理机制专门用于响应用户的操作，比如想要响应用户的单击鼠标、按下键盘等操作，就需要使用 Android 的事件处理机制。

首先我们介绍几个概念。

- ◆ 事件对象（Event）：通常就是外部用户的一次操作，是在 Android 组件上发生的特定事件。
- ◆ 事件源（组件）：事件发生的场所，通常就是产生事件的组件。
- ◆ 事件监听器（Listener）：负责监听事件源上发生的事件，并对各种事件做出相应处理的

◆ 事件处理器：事件处理器对象负责对接收的事件进行对应处理。

事件对象、事件源、事件监听器、事件处理器在整个事件处理机制中都起着非常重要的作用，它们彼此之间有着非常紧密的联系。下面介绍事件处理的工作流程：首先为事件源注册事件监听器对象，当用户进行一些操作时，如按下鼠标或者释放键盘等，这些动作会触发相应的事件，如果事件源注册了事件监听器，将产生并传递事件对象，事件监听器接收事件对象，并对事件进行处理。

在程序中，如果想实现事件的监听机制，首先需要在 Activity 类中实现事件监听器的接口，例如，Activity 需要实现 View.OnClickListener、View.OnLongClickListener，如图 3-23 所示。

```
public class MainActivity extends AppCompatActivity
        implements View.OnClickListener, View.OnLongClickListener{
    protected void onCreate(Bundle savedInstanceState) {
        setContentView(R.layout.activity_main);
        initView();
    }
}
```

图 3-23　Activity 类中实现事件监听器的接口

接着通过 setOnClickListener 方法为事件源注册事件监听器对象，当事件源上发生事件时，便会触发事件监听器对象，如图 3-24 所示。

```
private Button mBtnTest1;
private Button mBtnTest2;
private void initView() {
    mBtnTest1 = (Button) findViewById(R.id.btn_test1);//变量初始化，指向 Button1
    mBtnTest1.setOnClickListener(this);//按钮单击事件源注册到 Activity 事件监听器
    mBtnTest1.setOnLongClickListener(this); //按钮双击事件注册到 Activity 监听器
    mBtnTest2 = (Button) findViewById(R.id.btn_test2);
    mBtnTest2.setOnClickListener(this);
    mBtnTest2.setOnLongClickListener(this);
}
```

图 3-24　通过 setOnClickListener 方法为事件源注册事件监听器对象

由事件监听器调用相应的方法来处理相应的事件；在 Activitity 中重写 onClick 方法来实现按钮的单击事件处理；根据视图的 id 判断哪个按钮被单击；使用 Toast 将按钮的信息显示出来，如图 3-25 所示。

```
public void onClick(View v) {
    switch (v.getId()) {
        case R.id.btn_test1:
            Toast.makeText(this, "按钮" + ((Button) v).getText().toString() + "被单击
                了",Toast.LENGTH_SHORT).show();
            break;
        case R.id.btn_test2:
            Toast.makeText(this, "按钮" + ((Button) v).getText().toString() + "被单击
                了",Toast.LENGTH_SHORT).show();
            break;
    }
}
```

图 3-25　Activity 类中重写 onClick 方法（单击事件处理）

由事件监听器调用相应的方法来处理相应的事件；在 Activitity 中重写 onLongClick 方法来实现按钮的长按事件处理；根据视图的 id 判断哪个按钮被单击；使用 Toast 将按钮的信息

显示出来，如图 3-26 所示。

```java
public void onLongClick(View v) {
    switch (v.getId()) {
        case R.id.btn_test1:
            Toast.makeText(this, "按钮" + ((Button) v).getText().toString() + "被单击
                了",Toast.LENGTH_SHORT).show();//Toast 界面提示
            break;
        case R.id.btn_test2:
            Toast.makeText(this, "按钮" + ((Button) v).getText().toString() + "被单击
                了",Toast.LENGTH_SHORT).show();
            break;
                            default:
                                break;
    }
    return true;
}
```

图 3-26　Activity 类中重写 onLongClick 方法（长按事件处理）

3.2.3　EditText

编辑框和进度条
组件（实践案例）

EditText 在开发中也是经常使用的而且比较重要的一个控件，它是用户跟应用进行数据传输的窗口，比如实现一个登录界面，需要用户输入账号和密码，然后我们开发者获取到用户输入的内容，提交给后台服务器进行判断再做相应的处理。如图 3-27 所示的是 EditText 的一个实例；在 EditText 输入框中输入内容，单击"获取输入内容"按钮，提示输入的内容信息。

图 3-27　EditText 的一个运行实例

首先我们设置 EditText 组件的布局属性，如图 3-28 所示。

```xml
<EditText
    android:id="@+id/editText"
    android:layout_width="wrap_content " //视图宽度
    android:layout_height="wrap_content " //视图高度
    android:layout_marginTop="32dp " //视图外边缘距离
    android:background="@drawable/bg_edittext" //视图背景
    android:gravity="center" //视图文字居中
    android:hint="请输入内容"//视图提示内容
```

图 3-28　EditText 组件的布局属性

```
android:inputType="textMultiLine" //输入内容为多行，可选
android:maxLines="2" //视图最大输入行数
android:padding="16dp" //视图文字离视图内边缘距离
android:textSize="20dp" //视图文字大小
app:layout_constraintEnd_toEndOf="parent" //视图右边界
app:layout_constraintStart_toStartOf="parent"//视图左边界
app:layout_constraintTop_toTopOf="parent" //视图上边界
/>
```

图 3-28　EditText 组件的布局属性（续）

EditText 的组件使用了基于 selector 的 XML 布局文件 bg_edittext 作为背景文件；存放的路径为 res/drawable/bg_edittext.xml，其代码如图 3-29 所示。res/drawable/bg_edittext.xml 支持两种状态，分别为正常状态"@drawable/bg_edittext_normal"和获得焦点状态"@drawable/bg_edittext_focused"两种状态。

```
<?xml version="1.0" encoding="utf-8"?>
<selector xmlns:android="http://schemas.android.com/apk/res/android">
    <item android:state_window_focused="false" android:drawable="@drawable/bg_edittext_normal" /><!--正常状态-->
    <item android:state_focused="true" android:drawable="@drawable/bg_edittext_focused" /><!--获得焦点状态-->
</selector>
```

图 3-29　bg_edittext.xml 文件代码

正常状态下 res/drawable/bg_edittext_normal.xml 文件代码如图 3-30 所示；定义了<stroke>边框的颜色；<solid>线条的填充色；框四周圆弧的半径<corners android:radius="4dp" />。

```
<?xml version="1.0" encoding="utf-8"?>
<shape xmlns:android="http://schemas.android.com/apk/res/android">
    <stroke
        android:width="1dp"
        android:color="#BDC7D8" /><!--边框颜色-->
    <solid android:color="#FFFFFF" /><!--填充色-->
    <corners android:radius="4dp" />
</shape>
```

图 3-30　bg_edittext_normal.xml 文件代码

在获得焦点状态下 res/drawable/bg_edittext_focus.xml 文件代码如图 3-31 所示；获得焦点状态下<stroke>边框的颜色做出了修改；<solid>线条的填充色不变；框四周圆弧的半径也保持不变<corners android:radius="4dp" />

```
<?xml version="1.0" encoding="utf-8"?>
<shape xmlns:android="http://schemas.android.com/apk/res/android">
    <stroke
        android:width="1dp"
        android:color="#ff0000" /><!--获得焦点边框颜色改变-->
    <solid android:color="#FFFFFF" /><!--填充色-->
    <corners android:radius="4dp" />
</shape>
```

图 3-31　bg_edittext_focus.xml 文件代码

下面介绍 EditText 控件的内容获取。使用 mEditText.getText()获取文本内容，使用 toString()方法将其转换为字符串，使用 trim()将字符串前后的空格删除，使用 Toast 方法将 EditText 控件的文本内容显示出来，如图 3-32 所示。

```java
private void submit() {
    String editTextString = mEditText.getText().toString().trim();
    if (TextUtils.isEmpty(editTextString)) {
        Toast.makeText(this, "请输入内容", Toast.LENGTH_SHORT).show();
        return;
    }
    Toast.makeText(this, "获取文本: " + editTextString, Toast.LENGTH_SHORT).show();
}
```

图 3-32　EditText 控件的内容获取代码

EditText 控件案例的主视图实现代码如图 3-33 所示。

```java
package cn.edu.sziit.chapter3_edittext;
import android.os.Bundle;
import android.support.v7.app.AppCompatActivity;
import android.text.TextUtils;
import android.view.View;
import android.widget.Button;
import android.widget.EditText;
import android.widget.Toast;
public class MainActivity extends AppCompatActivity implements View.OnClickListener {
    private EditText mEditText;
    private Button mBtnTest;
    @Override
    protected void onCreate(Bundle savedInstanceState) {
        super.onCreate(savedInstanceState);
        setContentView(R.layout.activity_main);
        initView();
    }
    private void initView() {
        mEditText = (EditText) findViewById(R.id.editText);
        mBtnTest = (Button) findViewById(R.id.btn_test);
        mBtnTest.setOnClickListener(this);
    }
    @Override
    public void onClick(View v) {
        switch (v.getId()) {
            case R.id.btn_test:
                submit();
                break;
        }
    }
    private void submit() {
        // validate
        String editTextString = mEditText.getText().toString().trim();
        if (TextUtils.isEmpty(editTextString)) {
            Toast.makeText(this, "请输入内容", Toast.LENGTH_SHORT).show();
            return;
        }
        Toast.makeText(this, "获取文本: " + editTextString,
                Toast.LENGTH_SHORT).show();
    }
}
```

图 3-33　EditText 控件案例的主视图实现代码

3.2.4　ProgressBar

ProgressBar 在开发中也是经常使用的而且比较重要的一个控件，ProgressBar 的应用场景很多，比如用户登录时，后台发请求及等待服务器返回信息，这个时候会用到进度条；或者在进行一些比较耗时的操作时,需要等待一段较长的时间,这个时候如果没有提示,用户可能会以为手机死机了,这样会大大降低用户体验,所以在需要进行耗时操作的地方,添加上进度条,让用户知道当前的程序在执行中,也可以直观地告诉用户当前任务的执行进度。如图 3-34 所示的是 ProgressBar 的一个运行实例,单击"手动增加进度"按钮,进度条会每次增加 10%。单击"自动增加进度"按钮,进度条会依次从 10%增加到 100%。

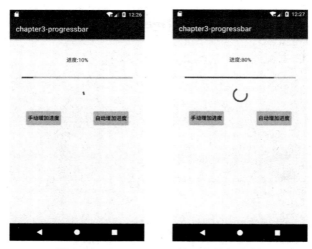

图 3-34　ProgressBar 的一个运行实例

首先我们设置一下 ProgressBar 组件的布局属性,ProgressBar 进度条的布局文件代码如图 3-35 所示。与左右的视图距离都是 32dp,在限制性属性中,顶部与进度文本控件对齐,与顶部的视图距离为 16dp。

```xml
<ProgressBar
    android:id="@+id/progressBar"
    style="?android:attr/progressBarStyleHorizontal"
    android:layout_width="0dp"
    android:layout_height="wrap_content"
    android:layout_marginEnd="32dp"
    android:layout_marginLeft="32dp"
    android:layout_marginRight="32dp"
    android:layout_marginStart="32dp"
    android:layout_marginTop="16dp"
    app:layout_constraintEnd_toEndOf="parent"
    app:layout_constraintStart_toStartOf="parent"
    app:layout_constraintTop_toBottomOf="@+id/textView" />
```

图 3-35　ProgressBar 进度条的布局文件代码

手动增加进度条功能代码如图 3-36 所示。进度变量 iProgress 每次增加 10%。mProgressBar 通过 setProgress 方法更改进度条的值,通过 mTextView 的 setText 方法设置进度文本控件的值。

```java
private void handIncrease() {
    iProgress += 10;
    if (iProgress >= 100) {
        iProgress = 0;
    }
    mProgressBar.setProgress(iProgress);
    mTextView.setText("进度:" + iProgress + "%");
}
```

图 3-36　手动增加进度条功能代码

自动增加进度条功能代码如图 3-37 所示。首先建立子线程，子线程中进度变量 iProgress 每次增加 10%，超过 100% 后重新归 0。子线程中操作主线程 UI，使用 runOnUiThread 在 UI 主线程空闲的时候设置 mProgressBar 进度条和 mTextView 文本控件，Thread.sleep 延时 200ms，使用 mThread.start() 启动线程。

```java
private void autoIncrease() {
    Thread mThread = new Thread(new Runnable() {
        @Override
        public void run() {
            while (iProgress <= 100) {
                iProgress += 10;
                if (iProgress > 100) {
                    iProgress = 0;
                    return;
                }
                runOnUiThread(new Runnable() {
                    @Override
                    public void run() {
                        mProgressBar.setProgress(iProgress);
                        mTextView.setText("进度:" + iProgress + "%");
                    }
                });
                try {
                    Thread.sleep(200);
                } catch (InterruptedException e) {
                    e.printStackTrace();
                }
            }
        }
    });
    mThread.start();
}
```

图 3-37　自动增加进度条功能代码

3.2.5　单线程模型

Android 开发过程中，常需要更新界面的 UI。而 UI 是要主线程（Main Thread）来更新的，即 UI 线程更新。如果在主线程之外的线程中直接更新页面显示常会报错，并且会抛出异常：android.view. ViewRoot$CalledFromWrongThreadException: Only the original thread that created a view hierarchy can touch its views。

当一个程序第一次启动时，Android 会同时启动一个对应的主线程，主线程主要负责处理与 UI 相关的事件，如用户的按键事件、用户接触屏幕的事件及屏幕绘图事件，并把相关的事件分发到对应的组件进行处理，所以主线程通常又被叫做 UI 线程。在开发 Android 应用时必须遵守单线程模型的原则：Android UI 操作并不是线程安全的并且这些操作必须在 UI 线程中执行。

子线程更新 UI：Android 的 UI 是单线程（Single-threaded）的。为了避免拖住 GUI，一些较费时的对象应该交给独立的线程去执行。如果由幕后的线程来执行 UI 对象，Android 就会发出错误信息 CalledFromWrongThreadException。

Message Queue：在单线程模型下，Android 设计了一个 Message Queue（消息队列），线程间可以通过该 Message Queue 并结合 Handler 和 Looper 组件进行信息交换。下面将对它们分别进行介绍。

（1）Message

Message（消息），理解为线程间交流的信息，处理数据后台线程需要更新 UI，则发送 Message（内含一些数据）给 UI 线程。

（2）Handler

Android 的控件是非线程安全的，在其他线程中操作主线程的控件将可能发生意外的情况，因此设计者希望对控件的操作依然交给主线程去完成，但在其他线程中，我们希望有一种机制去通知主线程来改变控件，因此 Handler 就诞生了，通过发送某个消息来告知主线程中的消息处理函数去操作控件。Handler（处理者）是 Message 的主要处理者，负责 Message 的发送和 Message 内容的执行处理。后台线程就是通过传进来的 Handler 对象引用来 sendMessage（Message）。而使用 Handler，需要 implement 该类的 handleMessage（Message）方法，它是处理这些 Message 的操作内容，例如 Update UI。通常需要子类化 Handler 来实现 handleMessage 方法。

（3）Message Queue

Message Queue（消息队列），用来存放通过 Handler 发布的消息，按照先进先出的原则执行。

每个 Message Queue 都会有一个对应的 Handler。Handler 会向 Message Queue 通过两种方法发送消息：sendMessage 或 post。这两种消息都会插在 Message Queue 队尾并按先进先出的原则执行。但通过这两种方法发送的消息执行的方式略有不同：通过 sendMessage 发送的是一个 message 对象，会被 Handler 的 handleMessage()函数处理；而通过 post 方法发送的是一个 runnable 对象，则会自己执行。

（4）Looper

Looper 是每条线程里的 Message Queue 的管家。Android 没有 Global 的 Message Queue，而 Android 会自动替主线程(UI 线程)建立 Message Queue，但在子线程里并没有建立 Message Queue。所以调用 Looper.getMainLooper()得到的主线程的 Looper 不为 NULL，但调用 Looper.myLooper()得到当前线程的 Looper 就有可能为 NULL。

Message 队列的工作原理如图 3-38 所示。

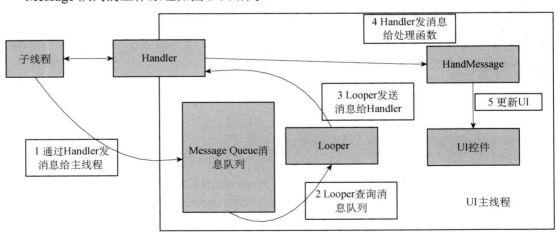

图 3-38　Message 队列的工作原理

Message 机制的流程如下：

① 在 Looper.loop()方法运行开始后，循环地按照接收顺序取出 Message Queue 里面的非

NULL 的 Message。

② 开始 Message Queue 里面的 Message 都是 NULL 的。当 Handler.sendMessage(Message) 到 Message Queue，该函数里面设置了 Message 对象的 target 属性是当前的 Handler 对象。随后 Looper 取出该 Message，则调用该 Message 中 target 所指向的 Handler 的 dispatchMessage 函数对 Message 进行处理。

在 dispatchMessage 方法中，如何处理 Message 则由用户指定，它有三个判断方法，优先级从高到低：

◆ Message 里面的 Callback，一个实现了 Runnable 接口的对象，其中 run 函数做处理工作。
◆ Handler 里面的 mCallback 指向的一个实现了 Callback 接口的对象，由其 handleMessage 进行处理。
◆ 处理消息 Handler 对象对应的类继承并实现了其中 handleMessage 函数，通过这个实现的 handleMessage 函数来处理消息。

③ Handler 处理完该 Message（update UI）后，Looper 则设置该 Message 为 NULL，以便回收。

非主 UI 线程更新视图的两种方法：第一种是 Handler；第二种是 Activity 中的 runOnUiThread（Runnable）方法。对于第一种方法，它采用的是传递消息的方式，调用 Handler 中的方法来处理消息更新视图。这种方法对于不是很频繁的调用是可取的。如果更新得较快，则消息处理会一直进行排队处理，这样显示会相对滞后。这个时候就可以考虑使用第二种方法，将需要执行的代码放到 Runnable 的 run 方法中，然后调用 runOnUiThread()这个方法将 Runnable 的对象传入即可。

在 ProgressBar 的实例中，实现了自动增加进度的功能，如图 3-39 所示的是 runOnUiThread 应用实例。runOnUiThread 可以替代 Handler 机制，并且更为灵活；自动增加进度的方法在线程中运用了 runOnUiThread 方法，runOnUiThread 方面运行了一个线程，这个线程会判断主线程是否空闲，如果主线程空闲就立刻更新组件的值；如果主线程不空闲的话，系统通过 Handler 机制将消息发送到主线程中，主线程空闲后再执行。

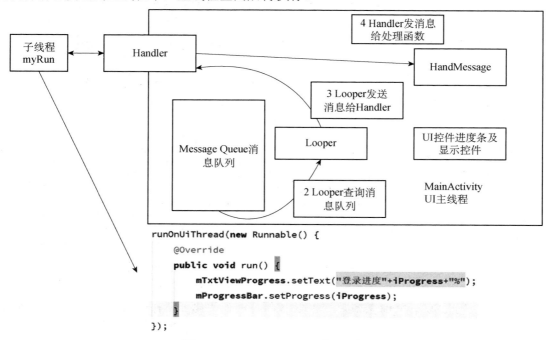

图 3-39　runOnUiThread 应用实例

3.2.6 单元小测

选择题：

1. 下面哪些组件表示文本？（　　）
A. ImageView　　　　B. EditText　　　　　　C. TextView　　　　　D. Button

2. 下面哪些组件表示图片？（　　）
A. ImageView　　　　B. EditText　　　　　　C. TextView　　　　　D. Button

3. 下面哪些组件表示文本输入框？（　　）
A. ImageView　　　　B. EditText　　　　　　C. TextView　　　　　D. Button

4. 下面哪些组件表示按钮？（　　）
A. ImageView　　　　B. EditText　　　　　　C. TextView　　　　　D. Button

5. 请问下面哪一个接口可以实现按钮单击事件？（　　）
A. View.OnClickListener
B. View.OnLongClickListener
C. AdapterView.OnItemClickListener
D. CompoundButton.OnCheckedChangeListener

6. 请问下面哪一个接口可以实现按钮长按事件？（　　）
A. View.OnClickListener
B. View.OnLongClickListener
C. AdapterView.OnItemClickListener
D. CompoundButton.OnCheckedChangeListener

7. 下面代码完成 EditText 控件文本的获取，请补全下面的代码？（　　）

```
private void submit() {
    String editTextString = （ ? ） ;
}
```

A. mEditText.getText()　　　　　　　　　B. mEditText.toString().trim()
C. mEditText.getText().toString().trim()　　D. mEditText.toString()

8. 下列代码用于设置进度条的进度，请补全代码？（　　）

```
private void handIncrease() {
    iProgress += 10;
    if (iProgress >= 100){
        iProgress = 0;
    }
    （?） ;
    mTextView.setText(""进度:"" + iProgress + ""%"");
}
```

A. mProgressBar.getProgress(iProgress);　　　B mProgressBar.setProgress();
C mProgressBar.getProgress();　　　　　　　D. mProgressBar.setProgress(iProgress);

9. 下面代码完成 Button1 按钮单击和长按事件的初始化，请补全代码？（　　）（多选）

```
private void initView() {
    mBtnTest1 = (Button) findViewById(R.id.btn_test1);
```

}
（ ？ ）

A. mBtnTest1.setOnClickListener(this);
B. setContentView(R.layout.activity_main);
C. mBtnTest1.setOnLongClickListener(this);
D. mBtnTest2.setOnClickListener(this);

3.3 Android 中级组件

本节我们主要介绍 Android 的中级组件，主要包括下面组件。
◆ CheckBox：复选框组件，主要介绍 CheckBox 的布局属性和使用方法。
◆ Switch：开关组件，主要介绍 Switch 的事件响应机制和使用方法。
◆ RadioButton：圆形单选框组件，主要介绍 RadioButton 的布局属性和使用方法。
◆ ImageView：图片组件，主要介绍 ImageView 的布局属性和使用方法。

中级组件（慕课）

3.3.1 CheckBox

CompoundButton 类是 Android 提供的抽象的复合按钮类，直接继承自 Button，它提供了具有两个状态的按钮：已选中或未选中；当按下或单击按钮时，状态会自动更改。由于 CompoundButton 是抽象类，因此在实际的开发中并不能直接使用，实际使用的是 CompoundButton 的几个派生类，主要有复选框 CheckBox 组件、开关按钮 Switch 组件、单选按钮 RadioButton 组件、ToggleButton 等，这些派生类都可以使用 CompoundButton 的属性和方法。

CompoundButton 相比较 Button 而言多出了一个监听事件接口 CompoundButton.OnCheckedChangeListener，当复合按钮的检查状态发生变化时调用。实现方法为：onCheckedChanged(CompoundButton buttonView,boolean isChecked)，其中参数 buttonView 表示复合按钮视图的状态；参数 isChecked 表示 buttonView 的新状态。

CompoundButton 提供了下面的方法对组件进行属性的设置。
◆ isChecked()：获取 Button 的当前状态。
◆ setButtonDrawable：设置按钮的勾选图标。
◆ setChecked(boolean checked)：更改按钮的状态。
◆ setOnCheckedChangeListener(CompoundButton.OnCheckedChangeListener listener)：设置按钮状态变化的监听器。

CheckBox 在开发中也是经常使用的而且比较重要的一个控件，它是一个复选框，是可以选中或取消选中的特定类型的双状态按钮。如图 3-40 所示的是 CheckBox 的一个运行实例；勾选 CheckBox 按钮，系统提示被勾选按钮的内容信息。

复选框和开关组件
（实践案例）

图 3-40 CheckBox 的一个运行实例

CheckBox 的自定义复选框使用了基于 selector 的 XML 布局文件 checkbox_selector.xml 作为背景文件；存放的路径为 res/drawable/checkbox_selector.xml，文件代码如图 3-41 所示。res/drawable/checkbox_selector.xml 支持两种状态，分别为被勾选状态"@drawable/check_choose"和不被勾选状态"@drawable/ check_unchoose"。

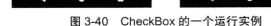

图 3-41 checkbox_selector.xml 文件代码

CheckBox 组件的初始化代码如图 3-42 所示。Activity 中实现复选框按钮单击事件监听器，使用了父类接口（CompoundButton.OnCheckedChangeListener）；定义 CheckBox 变量并初始化，将复选框选中事件注册到 Activity 事件监听器。

图 3-42 CheckBox 组件的初始化代码

CheckBox 的事件响应代码如图 3-43 所示。在 Activitity 中重写父类 CompoundButton 的

onCheckedChanged 方法以实现按钮的单击事件处理；onCheckedChanged 方法有两个参数，第一个参数 compoundButton 表示被单击的复选框；第二个参数 b 表示复选框按钮是否被选中。首先将父类组件 compoundButton 强制转换为 CheckBox 对象，根据 onCheckedChanged 传递的参数 b 判断复选框按钮是否被选中；使用 mCheckBox.getText()获取被选中组件的文本内容；使用 mCheckBox.getId()获取被选中组件的组件 ID；使用 mTextView.setText 将复选框信息显示出来。

```
public void onCheckedChanged(CompoundButton compoundButton, boolean b) {
    CheckBox mCheckBox = (CheckBox) compoundButton;
    String strMsg ="";
    if (b) {
        strMsg = String.format("控件:%s;ID:%d;被选中", mCheckBox.getText(), mCheckBox.getId());
    } else {
        strMsg = String.format("控件:%s;ID:%d;被取消", mCheckBox.getText(), mCheckBox.getId());
    }
    mTextView.setText(strMsg);
}
```

图 3-43 CheckBox 的事件响应代码

3.3.2 Switch

Switch 在开发中也是经常使用的而且比较重要的一个控件，它是一个开关组件，也是可以选中或取消选中的双状态按钮。如图 3-44 所示的是 Switch 的一个运行实例；滑动 Switch 按钮，系统提示被选中按钮的内容信息。

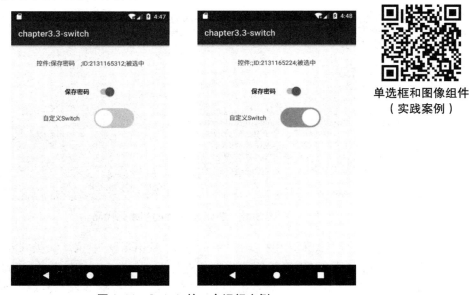

图 3-44 Switch 的一个运行实例

Switch 组件的 XML 属性列表和设置方法如表 3-7 所示。

表 3-7　Switch 组件的 XML 属性列表和设置方法

序号	XML 属性	Switch 类设置方法	说明
1	android:textOn	settextOn	控件打开显示的文字
2	android:textOff	settextOff	控件关闭显示的文字
3	android:switchPadding	setSwitchPadding	左右开关按钮的距离
4	android:thumb	setThumbDrawable	开关轨道背景

　　Switch 的自定义复选框使用了基于 selector 的 XML 布局文件 switch_selector.xml 作为背景文件,存放的路径为 res/drawable/switch_selector.xml,文件代码如图 3-45 所示。res/drawable/switch_selector.xml 支持两种状态,分别为被选中状态"@drawable/switch_on"和不被选中状态"@drawable/ switch_off"。

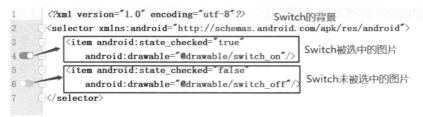

图 3-45　switch_selector.xml 文件代码

　　Switch 组件的初始化代码如图 3-46 所示：Activity 中实现复选框按钮单击事件监听器,使用了父类接口（CompoundButton.OnCheckedChangeListener）；定义 Switch 变量并初始化,将复选框选中事件注册到 Activity 事件监听器。

```
public class MainActivity extends AppCompatActivity implements CompoundButton.OnCheckedChangeListener{
    protected void onCreate(Bundle savedInstanceState) {
        setContentView(R.layout.activity_main);
        initView();
    }
}
private TextView mTextView;
private Switch mSwitch1;
private CheckBox mCheckBox1;
private void initView() {
    mTextView = (TextView) findViewById(R.id.textView);
    mSwitch1 = (Switch ) findViewById(R.id.switch1);
    mSwitch1.setOnCheckedChangeListener(this);
    mCheckBox1 = (CheckBox) findViewById(R.id.checkBox1);
    mCheckBox1.setOnCheckedChangeListener(this);
}
```

图 3-46　Switch 组件的初始化代码

　　Switch 的事件响应代码如图 3-47 所示；在 Activity 中重写父类 CompoundButton 的 onCheckedChanged 方法以实现按钮的单击事件处理；onCheckedChanged 方法有两个参数,第一个参数 compoundButton 表示被单击的复选框；第二个参数 b 表示复选框按钮是否被选中。首先将父类组件 compoundButton 强制转换为 Switch 对象；根据 onCheckedChanged 传递的参数 b 判断开关按钮是否被选中；使用 compoundButton.getText()获取被选中组件的文本内容；使用 compoundButton.getId()获取被选中组件的组件 ID；使用 mTextView.setText 将复选框信息显示出来。

```
public void onCheckedChanged(CompoundButton compoundButton, boolean b) {
    String strMsg = "";
    if (b) {
        strMsg = String.format("控件:%s;ID:%d;被选中", compoundButton.getText(),
            compoundButton.getId());
    } else {
        strMsg = String.format("控件:%s;ID:%d;被取消", compoundButton.getText(),
            compoundButton.getId());
    }
    mTextView.setText(strMsg);
}
```

图 3-47　Switch 的事件响应代码

3.3.3　RadioButton

RadioButton 在开发中也是经常使用的而且比较重要的一个控件，它是一个圆形的单选框组件。使用 RadioButton 必须和单选框 RadioGroup 一起使用，在 RadioGroup 中放置 RadioButton，用户在一个 RadioGroup 组中只能选择一个 RadioButton。如图 3-48 所示的是 RadioButton 的一个运行实例，在 RadioGroup 中选择 RadioButton，系统提示被选中 RadioButton 的内容信息。

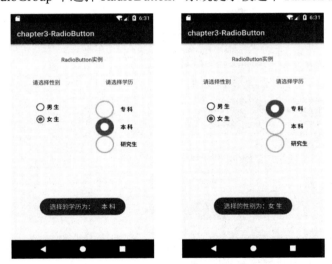

图 3-48　RadioButton 的一个运行实例

RadioButton 组件的 XML 属性列表和设置方法如表 3-8 所示。

表 3-8　RadioButton 组件的 XML 属性列表和设置方法

序号	XML 属性	RadioButton 类设置方法	说明
1	android:check	check	选中指定 ID 的按钮
2		getCheckedRadioButtonId	获取被选中按钮的 ID
3		setOnCheckedChangeListener	设置勾选状态的监听器

RadioButton 的自定义复选框使用了基于 selector 的 XML 布局文件 radio_selector.xml 作为背景文件；存放的路径为 res/drawable/radio_selector.xml，文件代码如图 3-49 所示；res/drawable/radio_selector.xml 支持两种状态，分别为被选中状态"@drawable/radio_choose"和不被选中

状态"@drawable/ radio_unchoose"。

```xml
<?xml version="1.0" encoding="utf-8"?>
<selector xmlns:android="http://schemas.android.com/apk/res/android">
    <item android:state_checked="true"
          android:drawable="@drawable/radio_choose"/>     → 按钮被选中状态
    <item android:state_checked="false"
          android:drawable="@drawable/radio_unchoose"/>   → 按钮未被选中状态
</selector>
```

图 3-49 radio_selector.xml 文件代码

RadioButton 实例的布局代码如图 3-50 所示。

```xml
<?xml version="1.0" encoding="utf-8"?>
<android.support.constraint.ConstraintLayout xmlns:android="http://schemas.android.com/apk/res/android"
    xmlns:app="http://schemas.android.com/apk/res-auto"
    xmlns:tools="http://schemas.android.com/tools"
    android:layout_width="match_parent"
    android:layout_height="match_parent"
    tools:context=".MainActivity">
    <RadioGroup
        android:id="@+id/radioGroup1"
        android:layout_width="wrap_content"
        android:layout_height="wrap_content"
        android:layout_marginEnd="8dp"
        android:layout_marginRight="8dp"
        android:layout_marginTop="16dp"
        app:layout_constraintEnd_toEndOf="parent"
        app:layout_constraintStart_toStartOf="@+id/guideline2"
        app:layout_constraintTop_toBottomOf="@+id/textView4">
        <RadioButton
            android:id="@+id/radioButton1"
            android:layout_width="wrap_content"
            android:layout_height="wrap_content"
            android:layout_marginTop="16dp"
            android:layout_weight="1"
            android:button="@null"
            android:drawableLeft="@drawable/radio_selector"
            android:text="    专 科" />

        <RadioButton
            android:id="@+id/radioButton2"
            android:layout_width="wrap_content"
            android:layout_height="wrap_content"
            android:layout_weight="1"
            android:button="@null"
            android:drawableLeft="@drawable/radio_selector"
            android:text="    本 科" />
        <RadioButton
            android:id="@+id/radioButton3"
            android:layout_width="wrap_content"
            android:layout_height="wrap_content"
            android:layout_weight="1"
            android:button="@null"
            android:drawableLeft="@drawable/radio_selector"
            android:text="    研究生" />
    </RadioGroup>
    <TextView
        android:id="@+id/textView6"
        android:layout_width="wrap_content"
        android:layout_height="wrap_content"
        android:layout_marginEnd="8dp"
        android:layout_marginLeft="16dp"
```

图 3-50 RadioButton 实例的布局代码

```xml
        android:layout_marginRight="8dp"
        android:layout_marginStart="16dp"
        android:layout_marginTop="16dp"
        android:padding="8dp"
        android:text="请选择学历"
        app:layout_constraintEnd_toEndOf="parent"
        app:layout_constraintStart_toStartOf="@+id/guideline2"
        app:layout_constraintTop_toBottomOf="@+id/textView" />
    <TextView
        android:id="@+id/textView"
        android:layout_width="wrap_content"
        android:layout_height="51dp"
        android:layout_marginTop="16dp"
        android:padding="16dp"
        android:text="RadioButton 实例"
        app:layout_constraintLeft_toLeftOf="parent"
        app:layout_constraintRight_toRightOf="parent"
        app:layout_constraintTop_toTopOf="parent" />
    <RadioGroup
        android:id="@+id/radioGroup2"
        android:layout_width="wrap_content"
        android:layout_height="wrap_content"
        android:layout_marginEnd="16dp"
        android:layout_marginLeft="16dp"
        android:layout_marginRight="16dp"
        android:layout_marginStart="16dp"
        android:layout_marginTop="16dp"
        app:layout_constraintEnd_toStartOf="@+id/guideline2"
        app:layout_constraintStart_toStartOf="parent"
        app:layout_constraintTop_toBottomOf="@+id/textView4">

        <RadioButton
            android:id="@+id/radioButton4"
            android:layout_width="wrap_content"
            android:layout_height="wrap_content"
            android:layout_marginTop="16dp"
            android:layout_weight="1"
            android:text="男 生" />
        <RadioButton
            android:id="@+id/radioButton5"
            android:layout_width="wrap_content"
            android:layout_height="wrap_content"
            android:layout_weight="1"
            android:text="女 生" />
    </RadioGroup>
    <TextView
        android:id="@+id/textView4"
        android:layout_width="wrap_content"
        android:layout_height="wrap_content"
        android:layout_marginEnd="16dp"
        android:layout_marginLeft="16dp"
        android:layout_marginRight="16dp"
        android:layout_marginStart="16dp"
        android:layout_marginTop="16dp"
        android:padding="8dp"
        android:text="请选择性别"
        app:layout_constraintEnd_toStartOf="@+id/guideline2"
        app:layout_constraintStart_toStartOf="parent"
        app:layout_constraintTop_toBottomOf="@+id/textView" />
    <android.support.constraint.Guideline
        android:id="@+id/guideline2"
        android:layout_width="wrap_content"
        android:layout_height="wrap_content"
        android:orientation="vertical"
        app:layout_constraintGuide_percent="0.5" />
</android.support.constraint.ConstraintLayout>
```

图 3-50 RadioButton 实例的布局代码（续）

RadioButton 组件的初始化如图 3-51 所示。Activity 中实现复选框单击事件监听器，使用了父类接口（RadioGroup.OnCheckedChangeListener）；定义 RadioButton 和 RadioGroup 变量并初始化；使用 RadioGroup 对象将复选框选中事件注册到 Activity 事件监听器；将第一个选项设置为默认选中。

```java
public class MainActivity extends AppCompatActivity    implements RadioGroup.OnCheckedChangeListener{
    protected void onCreate(Bundle savedInstanceState) {
        setContentView(R.layout.activity_main);
        initView();
    }
    private RadioGroup mRadioGroup1;private RadioButton mRadioButton1;
    private RadioButton mRadioButton2;private RadioButton mRadioButton3;
    private void initView() {
        mRadioButton1 = (RadioButton) findViewById(R.id.radioButton1);
        mRadioGroup1 = (RadioGroup) findViewById(R.id.radioGroup1);
        mRadioGroup1.setOnCheckedChangeListener(this);
        mRadioGroup1.check(R.id.radioButton1);
        ……
    }
}
```

图 3-51　RadioButton 组件的初始化

RadioButton 的事件响应代码如图 3-52 所示。在 Activitity 中重写父类 radioGroup 的 onCheckedChanged 方法以实现按钮的单击事件处理；onCheckedChanged 方法有两个参数，第一个参数 RadioGroup 表示所有的单选框；第二个参数 i 表示第 i 个单选框被选中；使用 radioGroup.getCheckedRadioButtonId()获取被选中组件的 ID；通过 ID 获取被选中组件对象；使用 mRadioButton.getText()获取被选中组件的文本内容；使用 Toast 将圆形单选框的选择信息显示出来。

```java
public void onCheckedChanged(RadioGroup radioGroup, int i) {
    String strMsg = "";
    RadioButton mRadioButton;
    mRadioButton=(RadioButton)
    findViewById(radioGroup.getCheckedRadioButtonId());//根据 Group 获取被选中组件
    if (radioGroup.getId() == R.id.radioGroup1) {
        strMsg = "选择的学历为："+mRadioButton.getText().toString();
    }
    if (radioGroup.getId() == R.id.radioGroup2) {
        strMsg = "选择的性别为："+mRadioButton.getText().toString();
    }
    Toast.makeText(this, strMsg, Toast.LENGTH_SHORT).show();
}
```

图 3-52　RadioButton 的事件响应代码

3.3.4　ImageView

ImageView 在开发中也是经常使用的而且比较重要的一个控件，它是一个图像组件，用来显示图片。如图 3-53 所示的是 ImageView 的一个运行实例。在 RadioGroup 中选择 RadioButton，根据图片的拉伸类型 scaleType 的不同，组件显示图片的方式也不同。

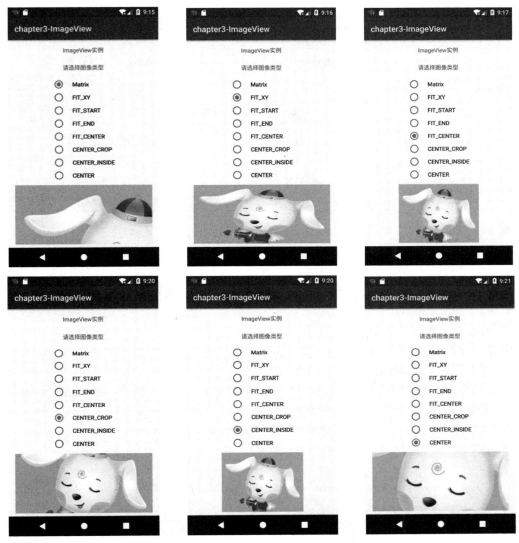

图 3-53 ImageView 的一个运行实例

ImageView 组件的 XML 属性列表和设置方法如表 3-9 所示。

表 3-9 ImageView 组件的 XML 属性列表和设置方法

序号	XML 属性	ImageView 类设置方法	说明
1	android:scaleType	setScaleType	设置图像的拉伸类型
2	android:src	setImageDrawable	设置图像 Drawable 对象
3	android:src	setImageResource	设置图像的资源 ID
4	android:src	setImageBitmap	设置图像的位图对象

android:scaleType 用于控制图片在组件的显示方式；scaleType 的基本属性和方法如表 3-10 所示。

表 3-10 scaleType 的基本属性和方法

序号	XML 属性	scaleType 拉伸类型值	说明
1	fitXY	FIT_XY	拉伸图片填满视图
2	fitStart	FIT_START	拉伸图片位于视图左方
3	fitCenter	FIT_CENTER	拉伸图片位于视图中心
4	fitEnd	FIT_END	拉伸图片位于视图右方
5	center	CENTER	保持图片原尺寸，并使其位于视图中间
6	centerCrop	CENTER_CROP	拉伸图片充满视图，并使其位于视图中间
7	centerInside	CENTER_INSIDE	图片位于视图中间（只压不拉）
8	matrix	MATRIX	动态缩小放大图片

- FIT_XY/fitXY：把图片按比例扩大/缩小到 View 的大小显示。
- FIT_START/fitStart：把图片按比例扩大/缩小到 View 的宽度，显示在 View 的上部分位置。
- FIT_CENTER/fitCenter：把图片按比例扩大/缩小到 View 的宽度，居中显示。
- FIT_END/fitEnd：把图片按比例扩大/缩小到 View 的宽度，显示在 View 的下部分位置。
- CENTER/center：按图片的原来 size 居中显示，当图片的长/宽超过 View 的长/宽，则截取图片的居中部分显示。
- CENTER_CROP/centerCrop：按比例扩大图片的 size 居中显示，使得图片的长（宽）等于或大于 View 的长（宽）。
- CENTER_INSIDE / centerInside：将图片的内容完整居中显示，通过按比例缩小或原来的 size 使得图片的长/宽等于或小于 View 的长/宽。
- MATRIX / matrix：用矩阵来绘制，动态缩小或放大图片来显示。

ImageView 实例的布局代码如图 3-54 所示。

```xml
<?xml version="1.0" encoding="utf-8"?>
<android.support.constraint.ConstraintLayout xmlns:android="http://schemas.android.com/apk/res/android"
    xmlns:app="http://schemas.android.com/apk/res-auto"
    xmlns:tools="http://schemas.android.com/tools"
    android:layout_width="match_parent"
    android:layout_height="match_parent"
    tools:context=".MainActivity">
    <RadioGroup
        android:id="@+id/radioGroup1"
        android:layout_width="wrap_content"
        android:layout_height="wrap_content"
        android:layout_marginEnd="16dp"
        android:layout_marginLeft="16dp"
        android:layout_marginRight="16dp"
        android:layout_marginStart="16dp"
        android:layout_marginTop="8dp"
        app:layout_constraintEnd_toEndOf="parent"
        app:layout_constraintStart_toStartOf="@+id/guideline2"
        app:layout_constraintTop_toBottomOf="@+id/textView6">
        <RadioButton
            android:id="@+id/radioButton1"
            android:layout_width="wrap_content"
            android:layout_height="wrap_content"
            android:layout_weight="1"
            android:checked="true"
            android:text="        Matrix" />
```

图 3-54 ImageView 实例的布局代码

```xml
<RadioButton
    android:id="@+id/radioButton2"
    android:layout_width="wrap_content"
    android:layout_height="wrap_content"
    android:layout_weight="1"
    android:text="      FIT_XY " />
<RadioButton
    android:id="@+id/radioButton3"
    android:layout_width="wrap_content"
    android:layout_height="wrap_content"
    android:layout_marginTop="0dp"
    android:layout_weight="1"
    android:text="      FIT_START" />
<RadioButton
    android:id="@+id/radioButton4"
    android:layout_width="wrap_content"
    android:layout_height="wrap_content"
    android:layout_weight="1"
    android:text="      FIT_END" />
<RadioButton
    android:id="@+id/radioButton5"
    android:layout_width="wrap_content"
    android:layout_height="wrap_content"
    android:layout_weight="1"
    android:checked="false"
    android:text="      FIT_CENTER " />
<RadioButton
    android:id="@+id/radioButton6"
    android:layout_width="wrap_content"
    android:layout_height="wrap_content"
    android:layout_weight="1"
    android:text="      CENTER_CROP" />
<RadioButton
    android:id="@+id/radioButton7"
    android:layout_width="wrap_content"
    android:layout_height="wrap_content"
    android:layout_weight="1"
    android:text="      CENTER_INSIDE" />
<RadioButton
    android:id="@+id/radioButton8"
    android:layout_width="wrap_content"
    android:layout_height="wrap_content"
    android:layout_weight="1"
    android:text="      CENTER " />
</RadioGroup>
<TextView
    android:id="@+id/textView6"
    android:layout_width="wrap_content"
    android:layout_height="wrap_content"
    android:layout_margin="8dp"
    android:layout_marginEnd="16dp"
    android:layout_marginLeft="16dp"
    android:layout_marginRight="16dp"
    android:layout_marginStart="16dp"
    android:layout_marginTop="16dp"
    android:padding="8dp"
    android:text="请选择图像类型"
    app:layout_constraintEnd_toEndOf="parent"
    app:layout_constraintStart_toStartOf="@+id/guideline2"
    app:layout_constraintTop_toBottomOf="@+id/textView" />
<TextView
    android:id="@+id/textView"
    android:layout_width="wrap_content"
    android:layout_height="wrap_content"
    android:layout_marginTop="8dp"
```

图 3-54 ImageView 实例的布局代码（续）

```xml
            android:padding="8dp"
            android:text="ImageView 实例"
            app:layout_constraintLeft_toLeftOf="parent"
            app:layout_constraintRight_toRightOf="parent"
            app:layout_constraintTop_toTopOf="parent" />
        <android.support.constraint.Guideline
            android:id="@+id/guideline2"
            android:layout_width="wrap_content"
            android:layout_height="wrap_content"
            android:orientation="vertical"
            app:layout_constraintGuide_percent="0" />
        <ImageView
            android:id="@+id/imageView"
            android:layout_width="0dp"
            android:layout_height="0dp"
            android:layout_marginBottom="8dp"
            android:layout_marginEnd="16dp"
            android:layout_marginLeft="16dp"
            android:layout_marginRight="16dp"
            android:layout_marginStart="16dp"
            android:layout_marginTop="8dp"
            android:scaleType="matrix"
            android:src="@drawable/cartoon"
            app:layout_constraintBottom_toBottomOf="parent"
            app:layout_constraintEnd_toEndOf="parent"
            app:layout_constraintStart_toStartOf="@+id/guideline2"
            app:layout_constraintTop_toBottomOf="@+id/radioGroup1" />
</android.support.constraint.ConstraintLayout>
```

图 3-54　ImageView 实例的布局代码（续）

下面使用 RadioButton 单选框来实现 ImageView 的各种 scaleType 的显示效果，实现流程如下：Activity 中实现复选钮单击事件监听器，使用了父类接口（RadioGroup.OnCheckedChangeListener）；定义 ImageView 和 RadioGroup 变量并初始化；使用 RadioGroup 对象将复选框选中事件注册到 Activity 事件监听器。ImageView 初始化代码如图 3-55 所示。

```java
package cn.edu.sziit.chapter3_imageview;
import android.os.Bundle;
import android.support.v7.app.AppCompatActivity;
import android.widget.ImageView;
import android.widget.RadioGroup;
public class MainActivity extends AppCompatActivity implements RadioGroup.OnCheckedChangeListener {
    private ImageView mImageView;
    private RadioGroup mRadioGroup1;
    @Override
    protected void onCreate(Bundle savedInstanceState) {
        super.onCreate(savedInstanceState);
        setContentView(R.layout.activity_main);
        initView();
    }
    private void initView() {
        mImageView = (ImageView) findViewById(R.id.imageView);
        mImageView.setImageResource(R.drawable.cartoon);
        mRadioGroup1 = (RadioGroup) findViewById(R.id.radioGroup1);
        mRadioGroup1.setOnCheckedChangeListener(this);
    }
}
```

图 3-55　ImageView 初始化代码

在 Activitity 中重写父类 radioGroup 的 onCheckedChanged 方法以实现按钮的单击事件处理；onCheckedChanged 方法有两个参数，第一个参数 radioGroup 表示所有的单选框；第二个参数 i 表示第 i 个单选框被选中；根据参数 i 来决定选择的是哪一个 scaleType；根据选择的 scaleType 拉伸类型设置 ImageView 图片的拉伸类型。RadioButton 的事件响应流程如图 3-56 所示。

```java
@Override
public void onCheckedChanged(RadioGroup radioGroup, int i) {
    switch (i) {
        case R.id.radioButton1:
            mImageView.setScaleType(ImageView.ScaleType.MATRIX);
            break;
        case R.id.radioButton2:
            mImageView.setScaleType(ImageView.ScaleType.FIT_XY);
            break;
        case R.id.radioButton3:
            mImageView.setScaleType(ImageView.ScaleType.FIT_START);
            break;
        case R.id.radioButton4:
            mImageView.setScaleType(ImageView.ScaleType.FIT_END);
            break;
        case R.id.radioButton5:
            mImageView.setScaleType(ImageView.ScaleType.FIT_CENTER);
            break;
        case R.id.radioButton6:
            mImageView.setScaleType(ImageView.ScaleType.CENTER_CROP);
            break;
        case R.id.radioButton7:
            mImageView.setScaleType(ImageView.ScaleType.CENTER_INSIDE);
            break;
        case R.id.radioButton8:
            mImageView.setScaleType(ImageView.ScaleType.CENTER);
            break;
    }
}
```

图 3-56　RadioButton 的事件响应流程

3.3.5　单元小测

选择题：

1. 下面哪些组件表示开关组件？（　　）
A. CheckBox　　　　　　　　　　　B. Switch
C. RadioButton　　　　　　　　　　D. ImageView

2 下面哪些组件表示复选框？（　　）
A. CheckBox　　　　　　　　　　　B. Switch
C. RadioButton　　　　　　　　　　D. ImageView

3 下面哪些组件表示图片组件？（　　）
A. CheckBox　　　　　　　　　　　B. Switch
C. RadioButton　　　　　　　　　　D. ImageView

4 下面哪些组件表示圆形单选框？（　　）
A .CheckBox　　　　　　　　　　　B. Switch
C. RadioButton　　　　　　　　　　D. ImageView

5 请问下面哪一个接口实现复选框的状态改变事件？（   ）

A. View.OnClickListener

B. View.OnLongClickListener

C. AdapterView.OnItemClickListener

D. CompoundButton.OnCheckedChangeListener

6. 请问 CheckBox 的父类是哪一个？（   ）

A. View   B. CompoundButton

C. Button   D. Box

7. 完成 CheckBox 的状态改变功能监听，请补全下面的代码？（   ）

```
private void initView() {
    mCheckBox = (CheckBox)findViewById(R.id.checkBox);
        (  ?  )
}
```

A. mCheckBox.setOnCheckedChangeListener(this);

B. mCheckBox.setOnClickListener(this);

C. mCheckBox.setOnLongClickListener(this);

D. mCheckBox.setOnItemClickListener(this);

8. 请问下面的哪个函数可以实现复选框 CheckBox 的事件监听器？（   ）

A. OnItemClick   B. OnClick

C. onCheckedChanged   D. OnLongClick

9. 请问下面哪一个接口可以实现圆形单选框的状态改变事件？（   ）

A. View.OnLongClickListener

B. View.OnClickListener

C. CompoundButton.OnCheckedChangeListener

D. AdapterView.OnItemClickListener

10. 请问圆形单选框的父类是哪一个？（   ）

A. Box   B. Button   C. View   D. CompoundButton

11 完成圆形单选框的状态改变功能监听，请补全下面的代码？（   ）

```
private void initView() {
    mRadioButton1 = (RadioButton) findViewById(R.id.radioButton1);
    mRadioGroup1 = (RadioGroup) findViewById(R.id.radioGroup1);
        (  ?  )
}
```

A. mRadioGroup.setOnCheckedChangeListener(this);

B. mRadioGroup.setOnClickListener(this);

C. mRadioGroup.setOnLongClickListener(this);

D. mRadioGroup.setOnItemClickListener(this);

12. 请问下面的哪个函数可以实现圆形单选框的事件监听器？（   ）

A. OnClick   B. onCheckedChanged

C. OnItemClick   D. OnLongClick

13. 在 ImageView 的视图中，如何设置图片的类型才能拉伸图片填满视图？（   ）

A. FIT_CENTER　　　　　　　　　　B. FIT_XY
C. CENTER　　　　　　　　　　　　D. CENTER_INSIDE

14. 在 ImageView 的视图中，如何设置图片的类型才能拉伸图片位于视图中心？（　　）
A. FIT_CENTER　　　　　　　　　　B. FIT_XY
C. CENTER　　　　　　　　　　　　D. CENTER_INSIDE

15. 在 ImageView 的视图中，如何设置图片的类型才能保持原图尺寸位于视图中心？（　　）
A. FIT_CENTER　　　　　　　　　　B. FIT_XY
C. CENTER　　　　　　　　　　　　D. CENTER_INSIDE

16. 在 ImageView 的视图中，如何设置图片的类型才能使图片位于视图中间（只压不拉）？（　　）
A. FIT_CENTER　　　　　　　　　　B. FIT_XY
C. CENTER　　　　　　　　　　　　D. CENTER_INSIDE

3.4 Android 适配器

本节我们主要介绍 Android 的适配器及对应的组件，主要包括以下几个。
◆ ListView：列表组件，主要介绍 ListView 与 SimpleAdapter 显示数据的实例。
◆ 自定义 Adapter：主要介绍 ListView 使用自定义 Adapter 显示数据的实例。
◆ Adapter 优化：主要介绍 Adapter 优化的意义和实现方法。

系统适配器
（慕课）

3.4.1 Adapter 适配器

Android 有许多列表组件需要显示多条数据，比如 Spinner、ListView、GridView、ViewPager 等。Adapter 作为数据与列表组件的桥梁，主要作用有以下两点：对要进行显示的数据进行处理；通过与视图对象绑定将数据显示到视图对象。Adapter、数据源和列表组件的关系如图 3-57 所示。Adapter 主要负责将不同的数据源显示到列表组件中，比如从数据库或者网络中获取的 Cursor 数据，还有以 ArrayList 为代表的列表数据。列表类组件通过不同的 Adapter 适配器将不同类型的数据进行显示。

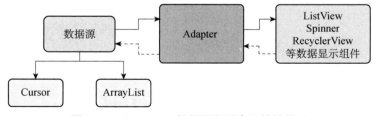

图 3-57　Adapter、数据源和列表组件的关系

Adapter 接口能把数据适配成 ListView 能访问的数据形式，Adapter 接口需要实现如图 3-58 所示的接口函数。从接口函数可以看出 Adapter 本身不维护数据，数据保存在数据存储区中（如 Array），但是 Adapter 适配了数据，如 getCount 返回数据的个数，getItem 返回指定的数据，同时 Adapter 还维护数据的显示，也就是 Item 子视图的显示，getView 需要返回一个 View 给 ListView。

```
Public interface Adapter
{
public abstract int getCount () //得到 listview 所要显示的总数目
public abstract object getItem(int i)//得到第 i 条条目对象
public abstract    View getView (int i, View view, ViewGroup parent)//得到当前条目的 view;
}
```

图 3-58　Adapter 接口需要实现的接口函数

Android 在适配器 Adapter 的基础上做了扩展来满足不同的需求，如图 3-59 所示的是 Android 适配器的层次体系；在我们使用过程中可以根据自己的需求实现接口或者继承类进行一定的扩展。

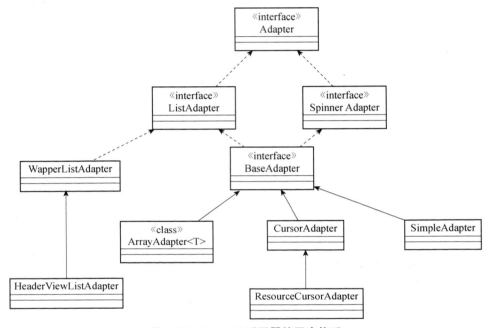

图 3-59　Android 适配器的层次体系

图 3-60　Spinner 的一个
运行实例

BaseAdapter 是一个抽象类，继承它需要实现较多的方法，所以也就具有较高的灵活性。

根据列表的适配器类型，列表分为三种：ArrayAdapter、SimpleAdapter 和 SimpleCursorAdapter，其中以 ArrayAdapter 最为简单，只能展示一行字。SimpleAdapter 有最好的扩充性，可以自定义出各种效果。SimpleCursorAdapter 可以认为是 SimpleAdapter 对数据库的简单结合，可以方便地把数据库的内容以列表的形式展示出来。

3.4.2　Spinner

Spinner 在开发中也是经常使用的而且比较重要的一个控件，其实它就是一个列表选择框。不过 Android 的列表选择框并不需要显示下拉列表，而是相当于弹出一个菜单供用户选择。如

系统适配器
（实践案例）

图 3-60 所示的是 Spinner 的一个运行实例；Spinner 显示了星期一到星期日的列表；在 Spinner 中选择子项，系统提示被选择的子项的内容信息。

Spinner 组件的 XML 属性列表和设置方法如表 3-11 所示。

表 3-11 Spinner 组件的 XML 属性列表和设置方法

序号	XML 属性	Spinner 类设置方法	说明
1	android:prompt	setPrompt	设置标题文字
2		setAdapter	设置下拉列表的适配器
3		setSelection	设置当前选择项
4		setOnItemSelectedListener	设置监听器

Spinner 实例的布局代码如图 3-61 所示。

```xml
<?xml version="1.0" encoding="utf-8"?>
<android.support.constraint.ConstraintLayout xmlns:android="http://schemas.android.com/apk/res/android"
    xmlns:app="http://schemas.android.com/apk/res-auto"
    xmlns:tools="http://schemas.android.com/tools"
    android:layout_width="match_parent"
    android:layout_height="match_parent"
    tools:context=".MainActivity">
    <TextView
        android:id="@+id/textView2"
        android:layout_width="wrap_content"
        android:layout_height="wrap_content"
        android:layout_marginTop="16dp"
        android:padding="8dp"
        android:text="Spinner 例子"
        app:layout_constraintLeft_toLeftOf="parent"
        app:layout_constraintRight_toRightOf="parent"
        app:layout_constraintTop_toTopOf="parent" />
    <Spinner
        android:id="@+id/spinner"
        android:layout_width="wrap_content"
        android:layout_height="wrap_content"
        android:layout_marginLeft="16dp"
        android:layout_marginRight="16dp"
        android:layout_marginTop="16dp"
        android:entries="@array/MyArray"
        android:padding="8dp"
        app:layout_constraintEnd_toEndOf="parent"
        app:layout_constraintStart_toStartOf="parent"
        app:layout_constraintTop_toBottomOf="@+id/textView2" />
</android.support.constraint.ConstraintLayout>
```

图 3-61 Spinner 实例的布局代码

Activity 中实现下拉列表框子项选中事件监听器的初始化如图 3-62 所示，使用了父类接口（AdapterView.OnItemSelectedListener），定义 Spinner 变量并初始化，将下拉列表框子项选中事件注册到 Activity 事件监听器。

```java
public class MainActivity extends AppCompatActivity
        implements AdapterView.OnItemSelectedListener{
    protected void onCreate(Bundle savedInstanceState) {
        setContentView(R.layout.activity_main);
        initView();
        initData();
    }
}
```

图 3-62 事件监听器的初始化代码

```
private TextView mTextView2;
private Spinner mSpinner;
private void initView() {
    mTextView2 = (TextView) findViewById(R.id.textView2);
    mSpinner = (Spinner) findViewById(R.id.spinner);
    mSpinner.setOnItemSelectedListener(this);
}
```

图 3-62 事件监听器的初始化代码(续)

数据初始化代码如图 3-63 所示。定义字符数组 String[]strWeek;定义并初始化 ArrayAdapter 对象,第一个参数代表当前的 Activity;第二个参数代表使用系统默认的布局显示 Spinner 的子项;第三个参数使用字符数组 strWeek;初始化 Spinner 组件对象;使用 setPrompt 设置标题;将 Spinner 组件对象与适配器对象绑定;使用 setSelection 设置当前的子项为第一行。

```
private String[] strWeek={"星期一","星期二","星期三","星期四","星期五","星期六","星期日"};
private void initData() {
    //初始化 ArrayAdapter
    ArrayAdapter<String> mArrayAdapter=new ArrayAdapter<String>
            (this,R.layout.support_simple_spinner_dropdown_item,strWeek);
    mSpinner.setPrompt("请选择日期");//设置标题
    mSpinner.setAdapter(mArrayAdapter);//组件与适配器绑定
    mSpinner.setSelection(0);//第一行
}
```

图 3-63 数据初始化代码

在 Activitity 中重写父类 AdapterView 的 onItemSelected 方法实现下拉列表框子项选中事件处理,如图 3-64 所示。ItemSelected 方法有三个参数,第一个参数 adapterView 表示 Spinner 组件对象;第二个参数 view 表示下拉框子项;第三个参数 i 代表第几个子项被选中;使用 Toast 将被选中的子项的信息进行显示。

```
public void onItemSelected(AdapterView<?> adapterView, View view, int i, long l) {
    String strMsg=String.format("选择了第%i 个选项:%s",i+1,strWeek[i]);
    Toast.makeText(this,strMsg,Toast.LENGTH_LONG).show();
}
```

图 3-64 onItemSelected 的事件响应流程

3.4.3 ListView

ListView 组件显示的是一个列表,也就是多个数据,那么这些数据从哪里获取,数据与 ListView 是如何关联起来的呢?一般来说,数据可以来自一个数组、一个 List,或者数据库中的游标。这些都是程序开发中常用到的数据集合的表示方式。但是 ListView 和数据本身并没有直接的关系,主要的原因是数据可以来自不同的方式。ListView 需要的数据可以是数据中的一个子集,另外 ListView 还需要显示每个子 View,而子 View 是不负责数据显示的,因此我们必须在数据和 ListView 中间增加一个第三者适配器 Adapter 来关联数据和 ListView,使得 ListView 不仅能获得数据,而且能把获取的数据以自己定制的方式呈现出来,并把这个呈现出来的数据转交给 ListView 来显示。Adapter 的作用就是把一些不兼容的、不能直接访问的数据适配成能兼容的、能访问的。在 Android 开发中 ListView 是比较常用的组件,它以列表的形式展示具体内容,并且能够根据数据的长度自适应显示,列表的显示需要三个元素,如图 3-65 所示。
- ListView:用来展示列表的视图 View。
- 适配器(Adapter):用来把数据映射到 ListView 上的中介。
- 数据源:包括具体的将被映射的字符串、图片,或者基本组件。

如图 3-66 所示的是 ListView 的一个运行实例。ListView 显示了校园风景的图片和说明;在 ListView 中选择子项,系统提示被选择子项的内容信息。

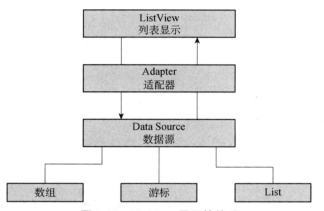

图 3-65　ListView 显示的关系　　　　　图 3-66　ListView 的一个运行实例

ListView 组件的 XML 属性列表和设置方法如表 3-12 所示。

表 3-12　ListView 组件的 XML 属性列表和设置方法

序号	XML 属性	Spinner 类设置方法	说明
1	android:divider	setDivider	设定分割线的图形
2	android:dividerHeight	setDividerHeight	设定分隔线的高度
3	android:headerDividersEnabled	setHeaderDividersEnabled	是否显示列表开头的分隔线
4	android:footerDividersEnabled	setFooterDividersEnabled	是否显示列表末尾的分隔线
5	android:divider	setDivider	设定分割线的图形

ListView 实例的布局代码如图 3-67 所示。

```xml
<?xml version="1.0" encoding="utf-8"?>
<android.support.constraint.ConstraintLayout xmlns:android="http://schemas.android.com/apk/res/android"
    xmlns:app="http://schemas.android.com/apk/res-auto"
    xmlns:tools="http://schemas.android.com/tools"
    android:layout_width="match_parent"
    android:layout_height="match_parent"
    tools:context=".MainActivity">
    <TextView
        android:id="@+id/textView2"
        android:layout_width="wrap_content"
        android:layout_height="wrap_content"
        android:layout_marginTop="16dp"
        android:padding="8dp"
        android:text="ListView 例子"
        app:layout_constraintLeft_toLeftOf="parent"
```

图 3-67　ListView 实例的布局代码

```
        app:layout_constraintRight_toRightOf="parent"
        app:layout_constraintTop_toTopOf="parent" />

<ListView
        android:id="@+id/listview"
        android:layout_width="0dp"
        android:layout_height="0dp"
        android:layout_margin="16dp"
        android:layout_marginEnd="8dp"
        android:layout_marginLeft="8dp"
        android:layout_marginRight="8dp"
        android:layout_marginStart="8dp"
        app:layout_constraintBottom_toBottomOf="parent"
        app:layout_constraintEnd_toEndOf="parent"
        app:layout_constraintStart_toStartOf="parent"
        app:layout_constraintTop_toBottomOf="@+id/textView2" />
</android.support.constraint.ConstraintLayout>
```

图 3-67　ListView 实例的布局代码（续）

ListView 一般使用 SimpleAdapter 作为适配器，SimpleAdapter 适配器主要用于显示 ListView 的数据；数据源可以为组合类型，ListView 组件通过 SimpleAdapter 将组合类型数据显示到组件；SimpleAdapter 的参数主要有以下 5 个：第一个参数为上下文对象；第二个参数为数据源，是含有 Map 的一个集合；第三个参数为每一个 item 的布局文件；第四个参数为 new String[]{}数组，数组中的每一项要与第二个参数中的存入 Map 集合的的 KEY 值一样，一一对应；第五个参数为 new int[]{}数组，存储的是第三个参数布局文件中的各个组件的 ID。

ListView 与 SimpleAdapter 适配器的原理图如图 3-68 所示。

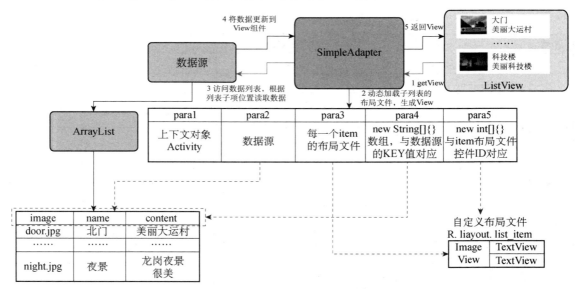

图 3-68　ListView 与 SimpleAdapter 适配器的原理图

◆ ListView 组件向 SimpleAdapter 发送获取子项目视图请求。
◆ SimpleAdapter 组件动态加载子项列表布局文件，并生成子视图 View。
◆ SimpleAdapter 组件访问数据列表，并根据列表子项位置读取对应的数据。
◆ 根据取到的数据更新子视图 View 的各个组件。
◆ 将子视图 View 返回给 ListView 组件，ListView 显示子项。

ListView 中每一个子项 list_item.xml 文件的布局代码如图 3-69 所示。

```xml
<?xml version="1.0" encoding="utf-8"?>
<LinearLayout xmlns:android="http://schemas.android.com/apk/res/android"
    android:layout_width="match_parent"
    android:layout_height="wrap_content"
    android:orientation="horizontal">
    <ImageView
        android:id="@+id/image1"
        android:layout_width="100dp"
        android:layout_height="100dp"
        android:src="@mipmap/ic_launcher"
        android:layout_margin="5dp"/>
    <LinearLayout
        android:id="@+id/ll2"
        android:layout_width="match_parent"
        android:layout_height="100dp"
        android:orientation="vertical"
        android:layout_marginTop="5dp"
        android:layout_marginLeft="10dp">
        <TextView
            android:id="@+id/text1"
            android:layout_width="wrap_content"
            android:layout_height="wrap_content"
            android:text="名字"
            android:textSize="20sp"
            android:layout_marginTop="10dp"/>
        <TextView
            android:id="@+id/text2"
            android:layout_width="wrap_content"
            android:layout_height="wrap_content"
            android:text="内容"
            android:textSize="16dp"
            android:layout_marginTop="10dp"/>
    </LinearLayout>
</LinearLayout>
```

图 3-69　ListView 实例子项 list_item.xml 文件的布局代码

Activity 中实现列表选中事件监听器的代码如图 3-70 所示，使用了父类接口（AdapterView.OnItemSelectedListener），定义 ListView 变量并初始化，将列表子项选中的事件注册到 Activity 事件监听器。

```java
public class MainActivity extends AppCompatActivity
        implements AdapterView.OnItemClickListener{
    protected void onCreate(Bundle savedInstanceState) {
        setContentView(R.layout.activity_main);
        initView();
        initData();
    }
    private TextView mTextView2;
    private ListView mListview;
    private void initView() {
        mTextView2 = (TextView) findViewById(R.id.textView2);
        mListview = (ListView) findViewById(R.id.listview);
        mListview.setOnItemClickListener(this);
    }
}
```

图 3-70　列表选中事件监听器的代码

数据初始化代码如图 3-71 所示：定义图片 id 数组 int[]iPic，图片名称字符数组 String[]strName，图片说明字符数组 String[]strContent；定义多列数据列表对象 ArrayList<Map<String,

Object>> mArrayList;将显示的数据对应添加到多列数据列表对象。

```
private int[] iPic = {R.drawable.door, R.drawable.lib, R.drawable.science, R.drawable.rest, R.drawable.lake, R.drawable.medium,
R.drawable.night};
private String[] strName = {"大门","图书馆","科技楼","宿舍楼","天鹅湖","体育场","夜景"};
private String[] strContent = {"美丽大运村","美丽图书馆","美丽科技楼","美丽宿舍楼",
"美丽天鹅湖","美丽体育场","美丽夜景"};
private ArrayList<Map<String,Object>> mArrayList;
private void initData() {
    mArrayList=new ArrayList<>();
    for (int i=0;i<iPic.length;i++)
    {
        Map<String,Object> map=new HashMap<>();
        map.put("image",iPic[i]);
        map.put("name",strName[i]);
        map.put("content",strContent[i]);
        mArrayList.add(map);
    }
}
```

图 3-71　数据初始化代码

SimpleAdapter 适配器初始化代码如图 3-72 所示:第一个参数为上下文对象 this;第二个参数为多列数据集合 mArrayList;第三个参数为每一个 item 布局文件使用的系统默认的 R.layout.list_item;第四个参数为 new String[]{}数组,数组中的每一项要与第二个参数中存入的 mArrayListkey 值一样,并且一一对应;第五个参数为 new int[]{}数组,存储的是第三个参数布局文件中的各个组件的 ID。

```
private void initData() {
    SimpleAdapter mSimpleAdapter=new SimpleAdapter (this,
mArrayList,
            R.layout.list_item,
            new String[]{"image","name","content"},
            new int[]{R.id.image1,R.id.text1,R.id.text2});
    mListview.setAdapter(mSimpleAdapter);
}
```

图 3-72　SimpleAdapter 适配器初始化代码

在 Activitity 中重写父类 AdapterView 的 onItemClick 方法实现下拉列表框子项选中事件处理,如图 3-73 所示;onItemClick 方法有三个参数,第一个参数 adapterView 表示 ListView 组件对象;第二个参数 View 表示 ListView 子项;第三个参数 i 代表第几个子项被选中;使用 Toast 将被选中的子项的信息进行显示。

```
public void onItemClick(AdapterView<?> adapterView, View view, int i, long l) {
    String strMsg=String.format("选择了第%d 个子选项:%s",i+1);
    Toast.makeText(this,strMsg,Toast.LENGTH_LONG).show();
}
```

图 3-73　onItemClick 的事件响应流程

自定义适配器
(慕课)

3.4.4　自定义 Adapter

本小节介绍自定义适配器,一般使用 BaseAdapter 作为父类实现自定义 Adapter,BaseAdapter 作为可以扩展的 Adapter,主要需要实现下面 4 个接口。

◆ public int getCount():获取适配器中的数据个数。

- public Object getItem(int position)：获取数据索引对应的数据项。
- public Object getItemId(int position)：获取数据索引 ID。
- public View getView(int, View, ViewGroup)：获取每一行 item 的显示内容。

ListView 可以使用自定义 BaseAdapter 适配器来进行数据的显示，自定义 BaseAdapter 的原理图如图 3-74 所示。

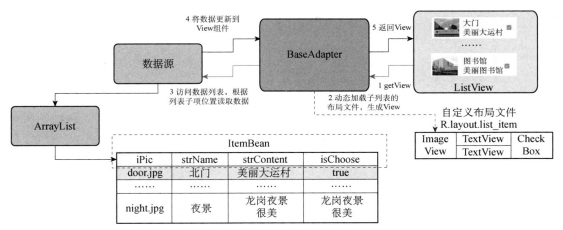

图 3-74　自定义 BaseAdapter 的原理图

- ListView 组件向 BaseAdapter 发送获取的子项目视图请求。
- BaseAdapter 组件动态加载子项列表布局文件，并生成子视图 View。
- BaseAdapter 组件访问数据列表，并根据列表子项位置读取对应的数据。
- 将数据更新到子视图 View 的各个组件。
- 将子视图 View 返回给 ListView 组件，ListView 显示子项。

自定义适配器
（实践案例）

ItemBean 的代码实现如图 3-75 所示；新建一个 ItemBean 类实现数据的聚合；定义图片 id、数据 int iPic、图片名称字符变量 String strName、图片说明字符变量 String strContent、图片是否被选中布尔变量 Boolean bChecked，定义变量的 Get 和 Set 方法。

```java
package cn.edu.sziit.chapter34_baseadapter;
public class ItemBean {
    private int iPic;
    private String strName;
    private String strContent;
    private Boolean bChecked;
    public int getiPic() {
        return iPic;
    }
    public void setiPic(int iPic) {
        this.iPic = iPic;
    }
    public String getStrName() {
        return strName;
    }
    public void setStrName(String strName) {
        this.strName = strName;
    }
    public String getStrContent() {
        return strContent;
```

图 3-75　ItemBean 的代码实现

```
        }
        public void setStrContent(String strContent) {
            this.strContent = strContent;
        }
        public Boolean getbChecked() {
            return bChecked;
        }
        public void setbChecked(Boolean bChecked) {
            this.bChecked = bChecked;
        }
}
```

图 3-75　ItemBean 的代码实现（续）

自定义 MyBaseAdapter 类继承 BaseAdapter，实现接口方法，如图 3-76 所示；定义上下文 Context mCtx 变量，定义 ArrayList<ItemBean> mArrayList 列表变量；以新定义的变量为参数定义 MyBaseAdapter 构造函数；实现 getCount 接口返回 mArrayList.size()；实现 getItem 接口返回 mArrayList.get(i)；实现 getItemId 接口返回位置参数 i；实现 getView 接口；将布局文件转化为 View 对象；定义并初始化 item 布局文件中对应的控件；获取子项位置的 ItemBean 数据对象；设置控件的对应属性值。

```java
package cn.edu.sziit.chapter34_baseadapter;
import android.content.Context;
import android.view.LayoutInflater;
import android.view.View;
import android.view.ViewGroup;
import android.widget.BaseAdapter;
import android.widget.CheckBox;
import android.widget.ImageView;
import android.widget.TextView;
import java.util.ArrayList;
public class MyBaseAdapter extends BaseAdapter {
    private Context mCtx;
    private ArrayList<ItemBean> mArrayList;
    public MyBaseAdapter(Context mCtx, ArrayList<ItemBean> mArrayList) {
        this.mCtx = mCtx;
        this.mArrayList = mArrayList;
    }
    @Override
    public int getCount() {
        return mArrayList.size();
    }
    @Override
    public Object getItem(int i) {
        return mArrayList.get(i);
    }
    @Override
    public long getItemId(int i) {
        return i;
    }
    @Override
    public View getView(int i, View view, ViewGroup viewGroup) {
        //将布局文件转化为 View 对象
        View mView = LayoutInflater.from(mCtx).inflate(R.layout.list_item, null);
        //找到 item 布局文件中对应的控件
        ImageView mImageView = (ImageView) mView.findViewById(R.id.image1);
        TextView mNameTextView = (TextView) mView.findViewById(R.id.text1);
        TextView mContentTextView = (TextView) mView.findViewById(R.id.text2);
        CheckBox mChekBox = (CheckBox) mView.findViewById(R.id.checkBox);
```

图 3-76　自定义 MyBaseAdapter 类的代码实现

```
        //获取相应索引的ItemBean对象
        ItemBean mItemBean = mArrayList.get(i);
        // 设置控件的对应属性值
        mImageView.setImageResource(mItemBean.getiPic());
        mNameTextView.setText(mItemBean.getStrName());
        mContentTextView.setText(mItemBean.getStrContent());
        if (mItemBean.getbChecked()) {
            mChekBox.setChecked(true);
        }
        return mView;
    }
}
```

图 3-76 自定义 MyBaseAdapter 类的代码实现（续）

数据初始化代码如图 3-77 所示：定义图片 ID 数组 int[]iPic，图片名称字符数组 String[] strName，图片说明字符数组 String[] strContent；定义多列数据列表对象 ArrayList<ItemBean> mArrayList；将显示的数据对应添加到多列数据列表对象。

```
private int[] iPic = {R.drawable.door, R.drawable.lib, R.drawable.science, R.drawable.rest,
        R.drawable.lake, R.drawable.medium, R.drawable.night};
private String[] strName = {"大门","图书馆","科技楼","宿舍楼","天鹅湖","体育场","夜景"};
private String[] strContent = {"美丽大运村","美丽图书馆","美丽科技楼","美丽宿舍楼",
"美丽天鹅湖","美丽体育场","美丽夜景"};
private ArrayList<ItemBean> mArrayList;
private void initData() {
    mArrayList=new ArrayList<>();
    for (int i=0;i<iPic.length;i++)
    {
        ItemBean mItemBean=new ItemBean();
        mItemBean.setiPic(iPic[i]);
            mItemBean.setStrName(strName[i]);
            mItemBean.setStrContent(strContent[i]);
            mItemBean.setbChecked(true);
            mArrayList.add(mItemBean);
    }
}
```

图 3-77 数据初始化代码

自定义适配器初始化代码如图 3-78 所示：定义 MyBaseAdapter 适配器对象并初始化，第一个参数使用当前 Activity 类 this；第二个参数使用数据列表对象 mArrayList；将 MyBaseAdapter 适配器对象设置为 ListView 对象的适配器。

```
private void initData() {
    MyBaseAdapter myBaseAdapter=new MyBaseAdapter(this,mArrayList);
            mListview.setAdapter(myBaseAdapter);
}
```

图 3-78 自定义适配器初始化代码

在 Activitity 中重写父类 AdapterView 的 onItemClick 方法实现下拉列表框子项选中事件处理，如图 3-79 所示；onItemClick 方法有三个参数，第一个参数 adapterView 表示 ListView 组件对象；第二个参数 view 表示 ListView 子项；第三个参数 i 代表第几个子项被选中；使用 Toast 将被选中的子项的信息进行显示；使用 mArrayList.get(i).getbChecked()获取子项的单选框状态，如果被选中就提示 CheckBox 被选中。

```
public void onItemClick(AdapterView<?> adapterView, View view, int i, long l) {
    String strMsg=String.format("选择了第%d 个子选项:%s",i+1);
    if (mArrayList.get(i).getbChecked())
    {
                        strMsg+="CheckBox 被选中";
    }
    Toast.makeText(this,strMsg,Toast.LENGTH_LONG).show();
}
```

图 3-79　利用 onItemClick 方法实现下拉列表子项选中事件处理

自定义的 Adapter 在实现 BaseAdapter 的 getView 方法的时候，主要存在下面的消耗资源的操作：ListView 的子视图由布局文件转化为 View 对象、占用内存、IO 操作影响性能；子视图的组件每次需要通过 findViewById 查找，比较耗时。

ListView 的 Adapter 的优化思路如图 3-80 所示。

假设屏幕每次只能显示 8 个子项 Item1～Item8；用户滑动屏幕的过程中，Item1 消失，增加 Item9；将 Item1 的视图对象通过 getView 参数 convertView 传递给 Item9；ListView 不管有多少子项，只需要准备 8 个 Item 对象内存。

生成 ViewHolder 对象，包括子布局的各个组件对象；Item 子布局生成并且初始化的过程中，将子布局的各个组件对象保存到 ViewHolder 对象中，并将 ViewHolder 对象存储到 Item 的 View 视图的 Tag 对象中；每次设置控件的对应属性值的时候，不需要通过 findViewById 查找子布局的控件，从而节省了耗时操作。

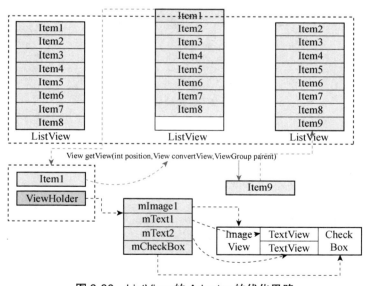

图 3-80　ListView 的 Adapter 的优化思路

ViewHolder 的实现代码如图 3-81 所示；在 getView 方法中选中子布局后单击 LayoutCreate 可以自动生成 ViewHolder 代码；ViewHolder 自动定义了属性变量，其中 View rootView 代表子布局视图；ImageView mImage1 代表图片组件；TextView mText1 代表图片名称；TextView mText2 代表图片说明；CheckBox mCheckBox 代表单选框；构造初始化函数将子布局的组件对象初始化。

```java
public static class ViewHolder {
    public View rootView;
    public ImageView mImage1;
    public TextView mText1;
    public TextView mText2;
    public CheckBox mCheckBox;
    public ViewHolder(View rootView) {
        this.rootView = rootView;
        this.mImage1 = (ImageView) rootView.findViewById(R.id.image1);
        this.mText1 = (TextView) rootView.findViewById(R.id.text1);
        this.mText2 = (TextView) rootView.findViewById(R.id.text2);
        this.mCheckBox = (CheckBox) rootView.findViewById(R.id.checkBox);
    }
}
```

图 3-81　ViewHolder 的实现代码

自定义 Adapter 类 MyBaseAdapter 中优化实现 getView 接口，如图 3-82 所示，主要的步骤如下：定义子视图 View 对象 mView 和子视图组件对象 mViewHolder；判断系统是否已提供子布局视图 view，如果提供，直接将系统提供的子布局视图 view 赋值给 mView，使用 view.getTag 方法赋值给 mViewHolder；如果系统不提供子布局视图，使用系统动态加载器将布局文件转换为 View 对象，并赋值给 mView；使用 mView 为参数生成 VIewHolder 对象将值赋值给 mViewHolder，使用 mView.setTag 方法将 mViewHolder 对象保存到 mView 视图中。

根据 getView 传递的参数子项位置 i，使用 mArrayList.get(i)获取对应子项的数据，使用 mViewHolder对象将子项的数据分别赋值给对应的组件对象，将子项视图mView传递回ListView。

```java
public View getView(int i, View view, ViewGroup viewGroup) {
    View mView;
    ViewHolder mViewHolder;
    if(view==null){//将布局文件转化为 View 对象
        mView=LayoutInflater.from(mCtx).inflate(R.layout.list_item, null);
        mViewHolder=new ViewHolder(mView);
        mView.setTag(mViewHolder);
    }
    else {
        mView=view;
        mViewHolder=(ViewHolder) mView.getTag();
    }//获取相应索引的 ItemBean 对象
    ItemBean mItemBean = mArrayList.get(i);
    mViewHolder.mImage1.setImageResource(mItemBean.getiPic());
    mViewHolder.mText1.setText(mItemBean.getStrName());
    mViewHolder.mText2.setText(mItemBean.getStrContent());
    if (mItemBean.getbChecked()) {
        mViewHolder.mCheckBox.setChecked(true);
    }
    return mView;
}
```

图 3-82　MyBaseAdapter 中优化实现 getView 接口

3.4.5　单元小测

选择题：

1. 请问组件与数据之间通过下面哪个类进行连接？（　　）
A. Adapter　　　　B. ListView　　　　C. Spinner　　　　D. GridView

2. 请问下面哪一个接口可以实现 Spinner 的下拉列表框选择事件？（　　）
A. View.OnClickListener
B. View.OnLongClickListener
C. AdapterView.OnItemClickListener
D. AdapterView.OnItemSelectedListener

3. 完成 Spinner 的下拉列表框选择事件监听，请补全下面的代码？（　　）

```
private void initView() {
    mSpinner = (Spinner) findViewById(R.id.spinner);
    (   ?   )
}
```

A. mSpinner.setOnItemSelectedListener(this)
B. mSpinner.setOnClickListener(this);
C. mSpinner.setOnLongClickListener(this);
D. mSpinner.setOnItemClickListener(this);

4. 请问下面的哪个函数可以实现 Spinner 的下拉列表框选择事件监听器？（　　）
A. onItemSelected　　B. OnClick　　C. OnItemClick　　D. OnLongClick

5. 请问下面哪一个接口可以实现 ListView 的列表单击事件？（　　）
A. View.OnClickListener
B. View.OnLongClickListener
C. AdapterView.OnItemSelectedListener
D. AdapterView.OnItemClickListener

6. 完成 ListView 的列表单击事件，请补全下面的代码。（　　）

```
private void initView() {
    mListview = (ListView) findViewById(R.id.listview);
    (   ?   )
}
```

A. mListview.setOnItemClickListener(this);
B mListview.setOnClickListener(this);
C. mListview.setOnLongClickListener(this);
D mListview.setOnItemSelectedListener(this)

7. 请问下面的哪个函数可以实现 ListView 的列表单击事件监听器？（　　）
A. OnItemClick　　B. OnClick　　C. onItemSelected　　D. OnLongClick

8. 请问下列哪些组件需要使用适配器显示数据？（　　）
A. Spinner　　B. RadioButton　　C. ViewPager　　D. ListView

本章课后练习和程序源代码

第 3 章源代码及课后习题

4 Android Fragment

 知识点

Android 的 Fragment 组件、ViewPager 组件。

 能力点

1. 熟练掌握 Android 的 Fragment 生命周期。
2. 熟练掌握 Android 的 Fragment 的加载和通信。
3. 熟练使用 Android 的 ViewPager 组件。

随着移动设备迅速发展，手机成为了生活必需品，平板电脑也变得越来越普及。一般手机屏幕的大小会在 4 英寸到 7 英寸之间，而平板电脑屏幕的大小会在 7 英寸到 10 英寸之间，屏幕大小的差距可能会让同样的界面在视觉效果上有较大的差异，比如一些界面在手机上看起来非常美观，但在平板电脑上看起来就可能会有控件被过分拉长，元素之间空隙过大。

Android 3.0（API 11）及以后的版本添加了一个强大的功能就是 Fragment（片段），主要是为了给大屏幕（如平板电脑）提供更加动态和灵活的 UI 设计支持。Fragment 能够同时兼顾手机和平板电脑的开发；可以让界面在平板电脑上更好地展示；Fragment 是能够嵌入到活动中的组件，可以将多个片段组合在一个 Activity 中来构建多窗格 UI，有自己的生命周期，并且可以有也可以没有用户界面。本章我们主要学习 Fragment 的使用方法。

 ## 4.1 Fragment 组件

Fragment 概述
（慕课）

碎片（Fragment）是一种可以嵌入到活动当中的 UI 片段，它能让程序更加合理和充分地利用大屏幕的空间，因而在平板电脑上应用得非常广泛。Fragment 与活动非常像，同样都能包含布局，同样都有自己的生命周期。那么如何使用碎片才能充分地利用平板电脑屏幕的空间呢？想象一下，我们正在开发一个新闻应用，其中一个界面使用 ListView 来展示一组新闻的标题，当单击了其中一个标题时，就打开另一个界面来显示新闻的详细内容。如果在手机中设计，我们可以将新闻标题列表放在一个活动中，将新闻的详细内容放在另一个活动中，如图 4-1 所示。

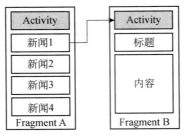

图 4-1 手机中显示新闻标题

可是如果在平板电脑上也这么设计，那么新闻标题列表将会被拉长至填充满整个平板电脑的屏幕，而新闻的标题一般都不会太长，这样将会导致界面上有大量的空白区域，如图 4-2 所示。

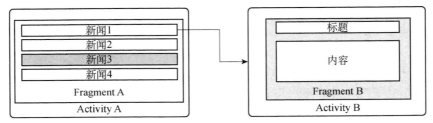

图 4-2 平板电脑中显示新闻标题

更好的设计方案是将新闻标题列表界面和新闻详细内容界面分别放在两个碎片中，然后在同一个活动中引入这两个碎片，就可以充分利用屏幕空间，如图 4-3 所示。

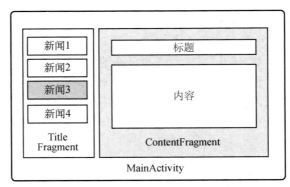

图 4-3 Fragment 显示新闻标题

4.1.1 Fragment 的生命周期

Fragment 片段必须始终嵌入在 Activity 中，其生命周期直接受宿主 Activity 生命周期的影响。例如，当 Activity 暂停时，其中的所有片

Fragment 静态加载
（实践案例）

段也会暂停；当 Activity 被销毁时，所有片段也会被销毁。不过，当 Activity 正在运行时，可以独立操纵每个片段，如添加或移除它们。当执行此类片段事务时，也可以将其添加到由 Activity 管理的返回栈当中。如图 4-4 所示的是一个 Fragment 显示新闻实例。主界面是一个 Activity，包含两个 Fragment，其中左边的 Fragment 包含了新闻 1 和新闻 2 两个按钮；右边的 Fragment 包含了一个新闻详细列表。

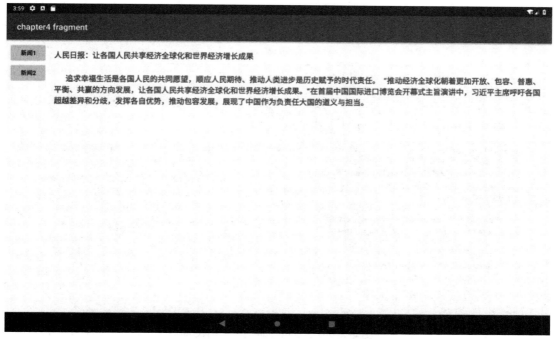

图 4-4 一个 Fragment 显示新闻实例

根据上面的 Fragment 显示新闻实例的图示，我们将 Fragment 布局分为三部分：主界面布局（activity_main.xml）、Title 布局（fragment_title.xml）、Content 内容布局（fragment_content.xml），如图 4-5 所示。

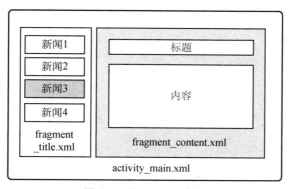

图 4-5 Fragment 布局

我们将 Fragment 代码实现分为三部分：主界面视图（MainActivity）、Title 视图（TitleFragment）、Content 内容视图（CotentFragment），如图 4-6 所示。

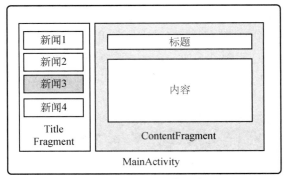

图 4-6　Fragment 代码实现

我们首先实现 Fragment 的布局，新建一个标题碎片布局 fragment_title.xml，代码如图 4-7 所示。fragment_title.xml 布局包含两个 Button。

```xml
<LinearLayout xmlns:android="http://schemas.android.com/apk/res/android"
    android:layout_width="match_parent"
    android:layout_height="match_parent"
    android:orientation="vertical">
    <Button
        android:id="@+id/btn_frag1"
        android:layout_width="match_parent"
        android:layout_height="wrap_content"
        android:text="新闻 1" />
    <Button
        android:id="@+id/btn_frag2"
        android:layout_width="match_parent"
        android:layout_height="wrap_content"
        android:text="新闻 2" />
</LinearLayout>
```

图 4-7　fragment_title.xml 布局代码

新建一个 TitleFragment，继承 android.support.v4.app.Fragment 类；TitleFragment 的代码实现如图 4-8 所示。首先重写 Fragment 的 onCreateView() 方法，然后在 onCreateView 方法中通过 LayoutInflater 的 inflate() 方法将刚才定义的 fragment_title.xml 布局动态加载进来；可以将 TitleFragment 视图的生命周期函数使用 Log 依次添加标签。

```java
package com.example.chapter4fragment;
import android.content.Context;
import android.os.Bundle;
import android.support.v4.app.Fragment;
import android.util.Log;
import android.view.LayoutInflater;
import android.view.View;
import android.view.ViewGroup;
import android.widget.Button;
public class TitleFragment extends Fragment implements View.OnClickListener {
    private Button mBtnFrag1;
    private Button mBtnFrag2;
    private String TAG="";
    @Override
    public View onCreateView(@NonNull LayoutInflater inflater, @Nullable ViewGroup container, @Nullable Bundle savedInstanceState) {
        View view = inflater.inflate(R.layout.fragment_title, container, false);
        initView(view);
        TAG=getClass().toString();
        Log.d(TAG, "onCreateView: ");
```

图 4-8　TitleFragment 的代码实现

```java
        return view;
    }
    @Override
    public void onAttach(Context context) {
        super.onAttach(context);
        Log.d(TAG, "onAttach: ");
    }
    @Override
    public void onCreate(@Nullable Bundle savedInstanceState) {
        super.onCreate(savedInstanceState);
        Log.d(TAG, "onCreate: ");
    }
    @Override
    public void onActivityCreated(@Nullable Bundle savedInstanceState) {
        super.onActivityCreated(savedInstanceState);
        Log.d(TAG, "onActivityCreated: ");
    }
    @Override
    public void onStart() {
        super.onStart();
        Log.d(TAG, "onStart: ");
    }
    @Override
    public void onResume() {
        super.onResume();
        Log.d(TAG, "onResume: ");
    }
    @Override
    public void onPause() {
        super.onPause();
        Log.d(TAG, "onPause: ");
    }
    @Override
    public void onStop() {
        super.onStop();
        Log.d(TAG, "onStop: ");
    }
    @Override
    public void onDestroyView() {
        super.onDestroyView();
        Log.d(TAG, "onDestroyView: ");
    }
    @Override
    public void onDestroy() {
        super.onDestroy();
        Log.d(TAG, "onDestroy: ");
    }
    @Override
    public void onDetach() {
        super.onDetach();
        Log.d(TAG, "onDetach: ");
    }
    private void initView(View view) {
        mBtnFrag1 = (Button) view.findViewById(R.id.btn_frag1);
        mBtnFrag2 = (Button) view.findViewById(R.id.btn_frag2);
        mBtnFrag1.setOnClickListener(this);
        mBtnFrag2.setOnClickListener(this);
    }
    @Override
    public void onClick(View v) {
        switch (v.getId()) {
            case R.id.btn_frag1:
                break;
            case R.id.btn_frag2:
                break;
        }
    }
}
```

图 4-8　TitleFragment 的代码实现（续）

新建一个标题碎片布局 fragment_content.xml，代码如图 4-9 所示。fragment_content.xml 布局包含两个 TextView。

```xml
<?xml version="1.0" encoding="utf-8"?>
<LinearLayout xmlns:android="http://schemas.android.com/apk/res/android"
    android:layout_width="match_parent"
    android:layout_height="match_parent"
    android:orientation="vertical">
    <TextView
        android:id="@+id/txt_title"
        android:layout_width="wrap_content"
        android:layout_height="wrap_content"
        android:layout_margin="8dp"
        android:gravity="center"
        android:padding="8dp"
        android:text="人民日报：让各国人民共享经济全球化和世界经济增长成果"
        android:textSize="18sp"
        android:textStyle="bold" />

    <TextView
        android:id="@+id/txt_content"
        android:layout_width="wrap_content"
        android:layout_height="wrap_content"
        android:layout_margin="8dp"
        android:layout_weight="1"
        android:gravity="top"
        android:padding="8dp"
        android:text="        追求幸福生活是各国人民的共同愿望，顺应人民期待、推动人类进步是历史赋予的时代责任。"
        android:textSize="18sp"
        android:textStyle="bold" />
</LinearLayout>
```

图 4-9　fragment_content.xml 布局

新建一个 ContentFragment，继承 android.support.v4.app.Fragment 类；ContentFragment 的代码实现如图 4-10 所示。首先重写 Fragment 的 onCreateView()方法，然后在 onCreateView 方法中通过 LayoutInflater 的 inflate()方法将刚才定义的 content_title.xml 布局动态加载进来；可以将 ContentFragment 视图的生命周期函数使用 Log 依次添加标签。

```java
package com.example.chapter4fragment;
import android.content.Context;
import android.os.Bundle;
import android.support.annotation.NonNull;
import android.support.annotation.Nullable;
import android.support.v4.app.Fragment;
import android.util.Log;
import android.view.LayoutInflater;
import android.view.View;
import android.view.ViewGroup;
import android.widget.TextView;
public class ContentFragment extends Fragment {
    private TextView mTxtTitle;
    private TextView mTxtContent;
    private String TAG="";
    @Nullable
    @Override
    public View onCreateView(@NonNull LayoutInflater inflater, @Nullable ViewGroup container, @Nullable Bundle savedInstanceState) {
        View view = inflater.inflate(R.layout.fragment_content, container, false);
        initView(view);
        TAG=getClass().toString();
```

图 4-10　ContentFragment 的代码实现

```java
        Log.d(TAG, "onCreateView: ");
        return view;
    }
    @Override
    public void onAttach(Context context) {
        super.onAttach(context);
        Log.d(TAG, "onAttach: ");
    }
    @Override
    public void onCreate(@Nullable Bundle savedInstanceState) {
        super.onCreate(savedInstanceState);
        Log.d(TAG, "onCreate: ");
    }
    @Override
    public void onActivityCreated(@Nullable Bundle savedInstanceState) {
        super.onActivityCreated(savedInstanceState);
        Log.d(TAG, "onActivityCreated: ");
    }
    @Override
    public void onStart() {
        super.onStart();
        Log.d(TAG, "onStart: ");
    }
    @Override
    public void onResume() {
        super.onResume();
        Log.d(TAG, "onResume: ");
    }
    @Override
    public void onPause() {
        super.onPause();
        Log.d(TAG, "onPause: ");
    }
    @Override
    public void onStop() {
        super.onStop();
        Log.d(TAG, "onStop: ");
    }
    @Override
    public void onDestroyView() {
        super.onDestroyView();
        Log.d(TAG, "onDestroyView: ");
    }
    @Override
    public void onDestroy() {
        super.onDestroy();
        Log.d(TAG, "onDestroy: ");
    }
    @Override
    public void onDetach() {
        super.onDetach();
        Log.d(TAG, "onDetach: ");
    }
    private void initView(View view) {
        mTxtTitle = (TextView) view.findViewById(R.id.txt_title);
        mTxtContent = (TextView) view.findViewById(R.id.txt_content);
    }
    public void setData(String strTitle,String strContent)
    {
        mTxtTitle.setText(strTitle);
        mTxtContent.setText(strContent);
    }
}
```

图 4-10　ContentFragment 的代码实现（续）

新建一个主视图布局 activity_main.xml，布局代码如图 4-11 所示：主视图布局包含

TitleFragment 和 ContentFragment 两个 Fragment 组件；对于 Fragment 的布局来说，必须在布局的时候指定布局对应的 Fragment 类，那么 TitleFragment 组件的实现类为 TitleFragment，ContentFragment 组件的实现类为 ContentFragment。

```xml
<?xml version="1.0" encoding="utf-8"?>
<android.support.constraint.ConstraintLayout xmlns:android="http://schemas.android.com/apk/res/android"
    xmlns:app="http://schemas.android.com/apk/res-auto"
    xmlns:tools="http://schemas.android.com/tools"
    android:layout_width="match_parent"
    android:layout_height="match_parent"
    tools:context=".MainActivity">
    <fragment
        android:id="@+id/fragment_title"
        android:name="com.example.chapter4fragment.TitleFragment"
        android:layout_width="0dp"
        android:layout_height="match_parent"
        android:layout_marginStart="8dp"
        android:layout_marginLeft="8dp"
        android:layout_marginTop="8dp"
        android:layout_marginBottom="8dp"
        app:layout_constraintBottom_toBottomOf="parent"
        app:layout_constraintStart_toStartOf="parent"
        app:layout_constraintTop_toTopOf="parent"></fragment>
    <fragment
        android:id="@+id/fragment_content"
        android:name="com.example.chapter4fragment.ContentFragment"
        android:layout_width="0dp"
        android:layout_height="match_parent"
        android:layout_marginTop="8dp"
        android:layout_marginEnd="8dp"
        android:layout_marginRight="8dp"
        android:layout_marginBottom="8dp"
        app:layout_constraintBottom_toBottomOf="parent"
        app:layout_constraintEnd_toEndOf="parent"
        app:layout_constraintStart_toEndOf="@+id/fragment_title"
        app:layout_constraintTop_toTopOf="parent"></fragment>
</android.support.constraint.ConstraintLayout>
```

图 4-11 activity_main.xml 布局代码

MainAcvitiy 主视图的代码实现如图 4-12 所示，将 MainActivity 视图的生命周期函数使用 Log 添加标签。

```java
package com.example.chapter4fragment;
import android.content.pm.ActivityInfo;
import android.os.Bundle;
import android.support.v7.app.AppCompatActivity;
import android.util.Log;
public class MainActivity extends AppCompatActivity {
    private String TAG="";
    @Override
    protected void onCreate(Bundle savedInstanceState) {
        super.onCreate(savedInstanceState);
        setContentView(R.layout.activity_main);
        TAG=getClass().toString();
        Log.d(TAG, "onCreate: ");
    }
    @Override
    protected void onStart() {
        super.onStart();
        Log.d(TAG, "onStart: ");
    }
```

图 4-12 MainActivity 主视图的代码实现

```java
@Override
protected void onResume() {
    /*** 设置为横屏*/
    if(getRequestedOrientation()!=ActivityInfo.SCREEN_ORIENTATION_LANDSCAPE){
        setRequestedOrientation(ActivityInfo.SCREEN_ORIENTATION_LANDSCAPE);
    }
    super.onResume();
    Log.d(TAG, "onResume: ");
}
@Override
protected void onPause() {
    super.onPause();
    Log.d(TAG, "onPause: ");
}
@Override
protected void onStop() {
    super.onStop();
    Log.d(TAG, "onStop: ");
}
@Override
protected void onDestroy() {
    super.onDestroy();
    Log.d(TAG, "onDestroy: ");
}
@Override
protected void onRestart() {
    super.onRestart();
    Log.d(TAG, "onRestart: ");
}
}
```

图 4-12　MainActivity 主视图的代码实现（续）

编写完程序后，Fragment 程序运行效果如图 4-13 所示。

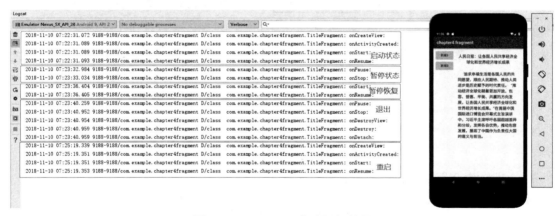

图 4-13　Fragment 程序运行效果

下面以 TitleFragment 为例，展示一下 Fragment 各种状态下的函数调用关系：

◆ Fragment 启动，经历 onCreateView->onActivityCreated->onStart->onResume 状态。

◆ Fragment 暂停，经历 onPause->onStop 状态。

◆ Fragment 暂停恢复，经历 onStart-> 和 onResume 状态。

◆ Fragmennt 退出，经历 onPause->onStop->onDestroy->onDetach 状态。

Frament 生命周期主要分为创建、开始、保持、暂停、停止、销毁几种状态，Frament 的生命周期如图 4-14 所示，生命周期图来自于 Android 的官网。

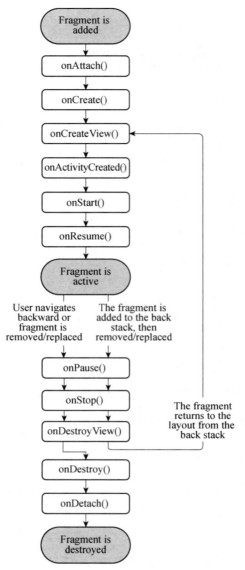

图 4-14　Fragment 的生命周期

Fragment 生命周期的各个函数的说明如下。

◆ onAttach：在片段与 Activity 关联之后调用。
◆ onCreate：在创建片段的时候调用。
◆ onCreateView：当为片段创建布局时调用，必须返回片段的根布局。如果片段未提供 UI，可以返回 NULL。
◆ onActivityCreated：告诉片段 Activity 的 onCreate()方法已经完成。
◆ onStart：当片段的视图对用户可见时调用。
◆ onResume：当 Activity 进入到 Resumed 状态的时候，也就意味着活动开始运行了，此时调用本方法。
◆ onPause：当 Activity 暂停的时候，调用该方法。系统将此方法作为用户离开片段的第一个信号（但并不总是意味着此片段会被销毁）进行调用。
- onStop：当 Activity 停止的时候调用。
- onDestoryView：调用本方法以允许片段释放用于视图的资源。
- onDestory：在片段销毁之前调用，以允许片段进行最好的清理工作。
- onDetach：当片段与 Activity 解除关联的时候调用。

Fragment 生命周期各种状态对应的函数调用流程图如图 4-15 所示。

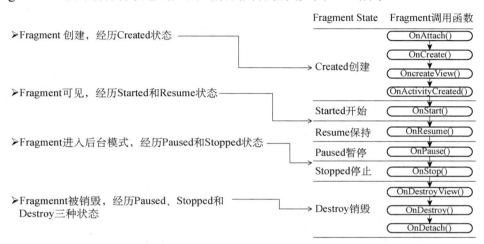

图 4-15　Fragment 生命周期各种状态对应的函数调用流程图

4.1.2　Fragment 通信

Fragment 与 Activity 通信（慕课）

本小节我们介绍 Fragment 之间的通信。虽然碎片都是嵌入在活动中显示的，可是实际上它们的关系并没有那么亲密。碎片和活动都是各自存在于一个独立的类当中的，它们之间并没有那么明显的方式来直接进行通信。如果想要在活动中调用碎片里的方法，或者在碎片中调用活动里的方法，应该如何实现呢？如图 4-16 所示是 Fragment 之间进行通信的示意图；为了方便碎片和活动之间进行通信，每个碎片中都可以通过调用 getActivity()方法来得到和当前碎片相关联的活动实例；Activity 中可以通过 FragmentManager 提供的类似于 findViewById()的方法，专门用于从布局文件中获取碎片的实例。

碎片之间如何进行通信呢？在上小节实现的新闻应用中，其中 TitleFragment 碎片使用 ListView 展示了一组新闻的标题，当单击其中一个标题，希望 ContentFragment 碎片显示新闻的详细内容。TitleFragment 通过 getActivity()获取关联的活动实例 MainActivity，MainActivity 通过 getSupportFragmentManager 方法获取碎片管理器，再通过 FragmentManager 提供的 findViewById()方法获取 ContentFragment 碎片实例，通过 ContentFragment 碎片实例提供的 setData 方法可以改变 ContentFragment 碎片显示新闻的详细内容，如图 4-16 所示。

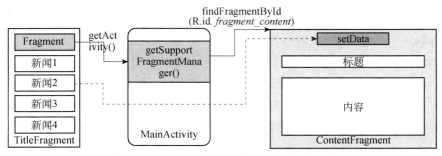

图 4-16 Fragment 之间进行通信的示意图

碎片之间通信的代码实现如图 4-17 所示；在 ContentFragment 碎片类中，新建一个 initView 的公共方法，方法中新建 mTxtTitle 和 mTxtContent 组件对象并初始化；新建一个公共的 setData 方法，参数分别为 String strTitle 和 String strContent 代表新闻标题和新闻内容，在方法中将 mTxtTitle 和 mTxtContent 组件对象分别设置为新闻标题和新闻内容。

```java
package com.example.chapter4fragment;
import android.content.Context;
import android.os.Bundle;
import android.support.annotation.NonNull;
import android.support.annotation.Nullable;
import android.support.v4.app.Fragment;
import android.util.Log;
import android.view.LayoutInflater;
import android.view.View;
import android.view.ViewGroup;
import android.widget.TextView;
public class ContentFragment extends Fragment {
    private TextView mTxtTitle;
    private TextView mTxtContent;
    private String TAG="";
    @Nullable
    @Override
    public View onCreateView(@NonNull LayoutInflater inflater, @Nullable ViewGroup container, @Nullable Bundle savedInstanceState) {
        View view = inflater.inflate(R.layout.fragment_content, container, false);
        initView(view);
        TAG=getClass().toString();
        Log.d(TAG, "onCreateView: ");
        return view;
    }
    @Override
    public void onAttach(Context context) {
        super.onAttach(context);
        Log.d(TAG, "onAttach: ");
    }
    @Override
    public void onCreate(@Nullable Bundle savedInstanceState) {
        super.onCreate(savedInstanceState);
        Log.d(TAG, "onCreate: ");
    }
    @Override
    public void onActivityCreated(@Nullable Bundle savedInstanceState) {
        super.onActivityCreated(savedInstanceState);
        Log.d(TAG, "onActivityCreated: ");
    }
    @Override
```

图 4-17 碎片之间通信的代码实现

```java
public void onStart() {
    super.onStart();
    Log.d(TAG, "onStart: ");
}
@Override
public void onResume() {
    super.onResume();
    Log.d(TAG, "onResume: ");
}
@Override
public void onPause() {
    super.onPause();
    Log.d(TAG, "onPause: ");
}
@Override
public void onStop() {
    super.onStop();
    Log.d(TAG, "onStop: ");
}
@Override
public void onDestroyView() {
    super.onDestroyView();
    Log.d(TAG, "onDestroyView: ");
}
@Override
public void onDestroy() {
    super.onDestroy();
    Log.d(TAG, "onDestroy: ");
}
@Override
public void onDetach() {
    super.onDetach();
    Log.d(TAG, "onDetach: ");
}
private void initView(View view) {
    mTxtTitle = (TextView) view.findViewById(R.id.txt_title);
    mTxtContent = (TextView) view.findViewById(R.id.txt_content);
}
public void setData(String strTitle,String strContent)
{
    mTxtTitle.setText(strTitle);
    mTxtContent.setText(strContent);
}
}
```

图 4-17　碎片之间通信的代码实现（续）

在 TitleFragment 碎片类中，新建一个 initView 的公共方法，新建 mBtnFrag 和 mBtnFrag2 两个 Button 对象并初始化，将两个 Button 对象设置监听。

新建一个公共的 setData 方法，在方法中使用通过 getActivity()获取关联的活动实例 MainActivity，MainActivity 通过 getSupportFragmentManager 方法获取碎片管理器，再通过 FragmentManager 提供的 findViewById()方法获取 ContentFragment 碎片实例，然后调用 ContentFragment 的 setData 方法。

重写 onClick 方法，根据事件响应的 Button 的 ID 号，调用 setData 方法设置新闻的标题和内容；TitleFragment 的代码实现如图 4-18 所示。

```java
package com.example.chapter4fragment;
import android.content.Context;
import android.os.Bundle;
import android.support.annotation.NonNull;
```

图 4-18　TitleFragment 的代码实现

```java
import android.support.annotation.Nullable;
import android.support.v4.app.Fragment;
import android.util.Log;
import android.view.LayoutInflater;
import android.view.View;
import android.view.ViewGroup;
import android.widget.Button;
public class TitleFragment extends Fragment implements View.OnClickListener {
    private Button mBtnFrag1;
    private Button mBtnFrag2;
    private String[] strTitle= {
            "人民日报：让各国人民共享经济全球化和世界经济增长成果",
            "美联储如期维持利率不变,明确进一步渐进加息"
    };
    private String[] strContent= {
            "    追求幸福生活是各国人民的共同愿望,顺应人民期待、推动人类进步是历史赋予的时代责任。", "    北京时间11月9日凌晨3时,美联储发布议息决议声明,维持联邦基金利率在2.00%-2.25%不变,符合市场预期。今年以来,美联储已经加息3次,分别在3月、6月和9月。此前市场预期,今年美联储还将加息一次,时间在12月。" };
    private String TAG="";
    @Nullable
    @Override
    public View onCreateView(@NonNull LayoutInflater inflater, @Nullable ViewGroup container, @Nullable Bundle savedInstanceState) {
        View view = inflater.inflate(R.layout.fragment_title, container, false);
        initView(view);
        TAG=getClass().toString();
        Log.d(TAG, "onCreateView: ");
        return view;
    }
    @Override
    public void onAttach(Context context) {
        super.onAttach(context);
        Log.d(TAG, "onAttach: ");
    }
    @Override
    public void onCreate(@Nullable Bundle savedInstanceState) {
        super.onCreate(savedInstanceState);
        Log.d(TAG, "onCreate: ");
    }
    @Override
    public void onActivityCreated(@Nullable Bundle savedInstanceState) {
        super.onActivityCreated(savedInstanceState);
        Log.d(TAG, "onActivityCreated: ");
    }
    @Override
    public void onStart() {
        super.onStart();
        Log.d(TAG, "onStart: ");
    }
    @Override
    public void onResume() {
        super.onResume();
        Log.d(TAG, "onResume: ");
    }
    @Override
    public void onPause() {
        super.onPause();
        Log.d(TAG, "onPause: ");
    }
    @Override
    public void onStop() {
        super.onStop();
        Log.d(TAG, "onStop: ");
    }
    @Override
    public void onDestroyView() {
```

图 4-18　TitleFragment 的代码实现（续）

```java
        super.onDestroyView();
        Log.d(TAG, "onDestroyView: ");
    }
    @Override
    public void onDestroy() {
        super.onDestroy();
        Log.d(TAG, "onDestroy: ");
    }
    @Override
    public void onDetach() {
        super.onDetach();
        Log.d(TAG, "onDetach: ");
    }
    private void initView(View view) {
        mBtnFrag1 = (Button) view.findViewById(R.id.btn_frag1);
        mBtnFrag2 = (Button) view.findViewById(R.id.btn_frag2);
        mBtnFrag1.setOnClickListener(this);
        mBtnFrag2.setOnClickListener(this);
    }
    @Override
    public void onClick(View v) {
        switch (v.getId()) {
            case R.id.btn_frag1:
                setContent(strTitle[0],strContent[0]);
                break;
            case R.id.btn_frag2:
                setContent(strTitle[1],strContent[1]);
                break;
        }
    }
    private void setContent(String strTitle, String strContent) {
        ContentFragment
mContentFragment=(ContentFragment)getActivity().getSupportFragmentManager().findFragmentById(R.id.fragment_content);
        mContentFragment.setData(strTitle,strContent);
    }
}
```

图 4-18 TitleFragment 的代码实现（续）

代码编写完成后在模拟器或者手机中运行代码，如图 4-19 所示的是碎片类通信的实现效果图；TitleFragment 碎片中单击"新闻 2"按钮，ContentFragment 碎片切换为新闻 2 的标题和新闻内容；单击"新闻 1"按钮，ContentFragment 碎片切换为新闻 1 的标题和新闻内容。

图 4-19 碎片类通信的实现效果图

4.1.3 Fragment 动态加载

在上一个新闻应用中，其中 TitleFragment 标题碎片和 ContentFragment 内容碎片在主视图布局中已固定，布局与碎片在布局的时候已一一对应，主视

Fragment 动态加载
（慕课）

Fragment 的动态加载
（实践案例）

图在加载 Fragment 碎片的过程中，根据 android:name 的属性碎片类的名字生成碎片实例，这种方式我们称为碎片的静态加载。

程序 App 主界面的常用例子如图 4-20 所示；App 下方有 4 个功能键，单击不同的功能键，希望能加载不同的碎片。

图 4-20　程序 App 主界面的常用例子——Fragment 动态加载实例

如果使用静态加载，上面的主视图布局只能加载一个碎片，不能实现动态切换的功能；这种情况下需要使用动态加载的功能，如图 4-21 所示，页面 1～页面 4 是 4 个不同的 Fragment，在功能键进行切换的时候，Activity 主视图动态加载对应的碎片进行切换。

图 4-21　Fragment 动态加载图

在主视图 activity_main.xml 布局中增加 FrameLayout，这是 Android 中最简单的一种布局，它没有任何的定位方式，所有的控件都会摆放在布局的左上角，FrameLayout 布局可以包含多个 Fragment，并根据要求动态加载指定的 Fragment；主视图 activity_main.xml 的整体结构代码如图 4-22 所示。

```xml
<?xml version="1.0" encoding="utf-8"?>
<LinearLayout xmlns:android="http://schemas.android.com/apk/res/android"
    xmlns:app="http://schemas.android.com/apk/res-auto"
    xmlns:tools="http://schemas.android.com/tools"
    android:layout_width="match_parent"
    android:layout_height="match_parent"
    android:orientation="vertical"
    tools:context=".MainActivity">
    <FrameLayout
        android:id="@+id/main_fm"
        android:layout_width="match_parent"
```

图 4-22　主视图 activity_main.xml 的整体结构代码

```xml
            android:layout_height="0dp"
            android:layout_weight="8">
        </FrameLayout>
        <LinearLayout
            android:layout_width="match_parent"
            android:layout_height="0dp"
            android:layout_weight="1"
            android:orientation="horizontal">
            <Button
                android:id="@+id/main_btn_1"
                android:layout_width="0dp"
                android:layout_height="match_parent"
                android:layout_weight="1"
                android:text="界面 1" />
            <Button
                android:id="@+id/main_btn_2"
                android:layout_width="0dp"
                android:layout_height="match_parent"
                android:layout_weight="1"
                android:text="界面 2" />
            <Button
                android:id="@+id/main_btn_3"
                android:layout_width="0dp"
                android:layout_height="match_parent"
                android:layout_weight="1"
                android:text="界面 3" />
            <Button
                android:id="@+id/main_btn_4"
                android:layout_width="0dp"
                android:layout_height="match_parent"
                android:layout_weight="1"
                android:text="界面 4" />
        </LinearLayout>
</LinearLayout>
```

图 4-22 主视图 activity_main.xml 的整体结构代码（续）

增加 Fragment1 到 Fragment4 的碎片布局，布局中只有一个 TextView，分别设置文本为页面 1 到页面 4。如图 4-23 所示的是 main_fragment_layout1.xml 布局代码。同理完成 main_fragment_layout2.xml、main_fragment_layout3.xml、main_fragment_layout4.xml 布局。

```xml
<?xml version="1.0" encoding="utf-8"?>
<LinearLayout xmlns:android="http://schemas.android.com/apk/res/android"
    android:layout_width="match_parent"
    android:layout_height="match_parent">
    <TextView
        android:id="@+id/textView"
        android:layout_width="match_parent"
        android:layout_height="wrap_content"
        android:layout_gravity="center_vertical|center_horizontal"
        android:gravity="center"
        android:text="页面 1"
        android:textSize="28dp" />
</LinearLayout>
```

图 4-23 main_fragment_layout1.xml 布局代码

为 Fragment1 到 Fragment4 的碎片布局新建 4 个 Fragment，继承 android.support.v4.app.Fragment 类；Fragment1 的代码实现如图 4-24 所示。首先重写 Fragment1 的 onCreateView() 方法，然后在 onCreateView() 方法中通过 LayoutInflater 的 inflate() 方法将刚才定义的 main_fragment_layout1.xml 布局动态加载进来；可以将 Fragment1 视图的生命周期函数使用 Log 依次添加标签。

```java
package cn.edu.sziit.fragmentdemo;
import android.content.Context;
import android.os.Bundle;
import android.support.annotation.Nullable;
import android.support.v4.app.Fragment;
import android.util.Log;
import android.view.LayoutInflater;
import android.view.View;
import android.view.ViewGroup;

public class Fragment1 extends Fragment {
    private String TAG="";
    @Override
    public View onCreateView(LayoutInflater inflater, ViewGroup container,
                             Bundle savedInstanceState) {
        TAG=getClass().toString();
        Log.d(TAG, "onCreateView: ");
        return inflater.inflate(R.layout.main_fragment_layout1,null);
    }
    @Override
    public void onAttach(Context context) {
        super.onAttach(context);
        Log.d(TAG, "onAttach: ");
    }
    @Override
    public void onCreate(@Nullable Bundle savedInstanceState) {
        super.onCreate(savedInstanceState);
        Log.d(TAG, "onCreate: ");
    }
    @Override
    public void onActivityCreated(@Nullable Bundle savedInstanceState) {
        super.onActivityCreated(savedInstanceState);
        Log.d(TAG, "onActivityCreated: ");
    }
    @Override
    public void onStart() {
        super.onStart();
        Log.d(TAG, "onStart: ");
    }
    @Override
    public void onResume() {
        super.onResume();
        Log.d(TAG, "onResume: ");
    }
    @Override
    public void onPause() {
        super.onPause();
        Log.d(TAG, "onPause: ");
    }
    @Override
    public void onStop() {
        super.onStop();
        Log.d(TAG, "onStop: ");
    }
    @Override
    public void onDestroyView() {
        super.onDestroyView();
        Log.d(TAG, "onDestroyView: ");
    }
    @Override
    public void onDestroy() {
        super.onDestroy();
        Log.d(TAG, "onDestroy: ");
    }
    @Override
    public void onDetach() {
```

图 4-24　Fragment1 的代码实现

```
            super.onDetach();
            Log.d(TAG, "onDetach: ");
        }
    }
```

图 4-24 Fragment1 的代码实现（续）

主视图实现碎片动态加载的代码如图 4-25 所示。增加一个 replaceFragment 方法，其中参数 int layout 代表主视图中的 FrameLayout 的资源 ID；Fragment fragment 代表需要实现的碎片实例。动态加载的流程如下：使用系统的 getSupportFragmentManager()获取碎片管理器；使用碎片管理器的 beginTransaction 获取事务管理；使用事务管理的 replace 方法替换当前的碎片视图；使用事务管理的 addToBackStack 方法将碎片加入碎片栈；使用事务管理的 commit 方法将此次碎片操作提交给主视图 Activity。

```
private void replaceFragment(int layout,Fragment fragment)
{
    FragmentManager manager=getSupportFragmentManager();
    FragmentTransaction   fragmentTransaction=manager.beginTransaction();
    fragmentTransaction.replace(layout,fragment);
    fragmentTransaction.addToBackStack(null);
    fragmentTransaction.commit();
}
```

图 4-25 主视图实现碎片动态加载的代码

主视图包含有 4 个按钮，根据视图的 getId 方法可以判断是哪一个按钮被按下；按钮被按下后调用 replaceFragment 方法，可以实现碎片的动态加载和切换，主视图的整体代码如图 4-26 所示。

```
package cn.edu.sziit.fragmentdemo;
import android.os.Bundle;
import android.support.v4.app.Fragment;
import android.support.v4.app.FragmentManager;
import android.support.v4.app.FragmentTransaction;
import android.support.v7.app.AppCompatActivity;
import android.util.Log;
import android.view.View;
import android.widget.Button;
import android.widget.FrameLayout;
public class MainActivity extends AppCompatActivity implements View.OnClickListener {
    private FrameLayout mMainFm;
    private Button mMainBtn1;
    private Button mMainBtn2;
    private Button mMainBtn3;
    private Button mMainBtn4;
    private String TAG="";
    @Override
    protected void onCreate(Bundle savedInstanceState) {
        super.onCreate(savedInstanceState);
        setContentView(R.layout.activity_main);
        initView();
        initFragment();
        TAG=getClass().toString();
        Log.d(TAG, "onCreate: ");
    }
    private void initFragment() {
        getSupportFragmentManager().beginTransaction().replace(R.id.main_fm,new Fragment1()).commit();
    }
    private void initView() {
        mMainFm = (FrameLayout) findViewById(R.id.main_fm);
```

图 4-26 主视图的整体代码

```
        mMainBtn1 = (Button) findViewById(R.id.main_btn_1);
        mMainBtn2 = (Button) findViewById(R.id.main_btn_2);
        mMainBtn3 = (Button) findViewById(R.id.main_btn_3);
        mMainBtn4 = (Button) findViewById(R.id.main_btn_4);
        mMainBtn1.setOnClickListener(this);
        mMainBtn2.setOnClickListener(this);
        mMainBtn3.setOnClickListener(this);
        mMainBtn4.setOnClickListener(this);
    }
    @Override
    public void onClick(View v) {
        switch (v.getId()) {
            case R.id.main_btn_1:
                replaceFragment(R.id.main_fm,new Fragment1());
                break;
            case R.id.main_btn_2:
                replaceFragment(R.id.main_fm,new Fragment2());
                break;
            case R.id.main_btn_3:
                replaceFragment(R.id.main_fm,new Fragment3());
                break;
            case R.id.main_btn_4:
                replaceFragment(R.id.main_fm,new Fragment4());
                break;
        }
    }
}
```

图 4-26　主视图的整体代码（续）

程序编写完成后在手机和模拟器上运行，程序运行的效果及页面的生命周期流程如图 4-27 所示。页面 1 启动后，页面跳转到页面 2，页面 2 返回后跳转到页面 1，页面 1 重启，页面 1 返回后退出。

图 4-27　程序运行的效果及页面的生命周期流程

4.1.4　单元小测

选择题：

1. 一个 Activity 可以启动多少个 Fragment？（　　　）

A. 2　　　　　　　　B. 1　　　　　　　　C. 多个　　　　　　　　D. 4

2. Fragment 类中最先启动的是哪一个方法？（　　）
A. onCreate()　　　B. onStart()　　　C. onCreateView()　　　D. onPause()

3 主视图布局中需要添加一个 Fragment，请补全下面的代码。（　　）

```
<fragment
android:id="@+id/fragment_content"
android:name="（   ?   ）"
</fragment>
```

A. ContentFragment　　B. TitleFragment　　C. MainAcitivty　　D. Activity

4. 主视图布局中需要添加一个 Fragment，请补全下面的代码。（　　）

```
<fragment android:id="@+id/fragment_title"
android:name="（   ?   ）"
</fragment>
```

A. ContentFragment　　B. TitleFragment　　C. MainAcitivty　　D. Activity

5. TitleFragment 中访问 ContentFragment 的方法，请补全下面的代码。（　　）

```
ContentFragment mContentFragment=(ContentFragment)getActivity().（   ?   ）.findFragmentById(R.id.fragment_content);
    mContentFragment.setData(strTitle,strContent);
```

A. getSupportManager()　　　　　　B. getManager()
C. getFrament()　　　　　　　　　D. getSupportFragmentManager()

6. Activity 中如果加载多个 Fragment，初始化显示第一个 Fragment，请补全下面的代码。（　　）

```
private void initFragment() {
    getSupportFragmentManager().（   ?   ）.replace(R.id.main_fm,new Fragment1()).commit();
}
```

A. Transaction()　　　　　　　　　B. begin()
C. beginTransaction()　　　　　　　D. getFrament()

7. Fragment 可以实现动态加载，下面是实现动态加载的过程，请补全代码。（　　）

```
private void replaceFragment(int layout,Fragment fragment){
    FragmentManager manager=getSupportFragmentManager();
    FragmentTransaction fragmentTransaction=manager.beginTransaction();
    fragmentTransaction.replace(layout,fragment);
    fragmentTransaction.addToBackStack(null);
    （   ?   ）;
}
```

A. fragmentTransaction.getFragmentById()　　B. fragmentTransaction.getAcitivity()
C. fragmentTransaction.commitFragment()　　D. fragmentTransaction.commit()

8. Activity 中如果加载多个 Fragment，切换显示第三个 Fragment，请问下面的代码哪些能实现？（　　）
A. replaceFragment(R.id.main_fm,new Fragment1())
B. replaceFragment(R.id.main_fm,new Fragment2())
C. replaceFragment(R.id.main_fm,new Fragment3())
D. replaceFragment(R.id.main_fm,new Fragment4())

9. Fragment 类中创建执行的方法是（   ）。
   A. onCreate()   B. OnAttach()   C. onStart()   D. onCreateView()
10. Fragment 碎片可见执行的方法是（   ）。
    A. onSaveInstance()   B. onStart()
    C. onResume()   D. onStop()
11. Fragment 进入后台模式执行的方法是（   ）。
    A. onRestart()   B. onPause()   C. onResume()   D. onStop()
12. Fragment 被销毁执行的方法是（   ）。
    A. onStop()   B. onDetachView() C. onDestroy()   D. onDetach()

4.2 ViewPager 组件

本节我们主要介绍 ViewPager 翻页类组件，主要介绍 ViewPager 控件加载 Fragment 的原理和适配器用法。

4.2.1 ViewPager 概述

ViewPager 翻页类组件（慕课）

ViewPager 视图翻页工具提供了多页面切换的效果。ViewPager 是 Android 3.0 后引入的一个 UI 控件，目前我们使用 Android Studio 进行开发，默认导入 v7 包；ViewPager 使用后，我们通过创建 adapter 给它填充多个 view，左右滑动时，切换不同的 view。Google 官方建议使用 Fragment 来填充 ViewPager，这样可以更加方便地生成每个碎片，以及管理每个碎片的生命周期。ViewPager 的运行实例如图 4-28 所示。滑动手机的屏幕，切换不同的 Fragment，展示出不同的校园风景图片。

图 4-28　ViewPager 的运行实例

ViewPager 基本属性和方法如表 4-1 所示。

表 4-1　ViewPager 基本属性和方法

序号	类设置方法	子项属性	说明
1	setAdapter		设置页面项的适配器，PagerAdapter 及其子类
2	setCurrentItem		设定当前页面的序号
3	addOnPageChangeListener：设置翻页视图的页面切换监听器	onPageScroll StateChanged	页面滑动状态变化时触发
		onPageScrolled	页面滑动过程中触发
		onPageSelected	选中页面时，滑动结束时触发

setAdapter：设置适配器；setCurrentItem：设置当前页面的序号；addOnPageChangeListener：设置组件的页面切换监听器；其中 onPageSelected 代表选中页面时，滑动结束时触发；onPageScrollStateChanged 代表页面滑动状态变化时触发；onPageScrolled 代表页面滑动过程中触发。

ViewPager 的原理图如图 4-29 所示：主视图中包含一个 ViewPager 组件；ViewPager 组件包含多个 Fragment 碎片组件。

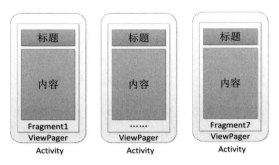

图 4-29　ViewPager 的原理图

ViewPager 的实现流程如图 4-30 所示，多个 Fragment 组成了一个 ArrayList 集合；ViewPager 与 Fragment 集合数据通过系统自定义的 FragmentPagerAdapter 进行适配。

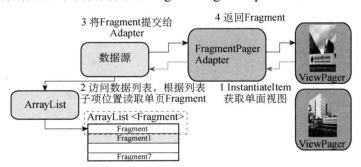

图 4-30　ViewPager 的实现流程

ViewPager 的实现流程如下：
◆ ViewPager 通过 FragmentPagerAdapter 适配器的 InstantiateItem 方法获取视图。
◆ 适配器访问数据源，根据组件的子项位置读取 Fragment 碎片。
◆ 数据源将指定位置的 Fragment 提交给适配器。

◆ 适配器将 Frament 对象返回给 ViewPager 组件显示。

首先实现主视图布局，主视图中 ViewPager 的布局代码如图 4-31 所示，ViewPager 的布局与普通组件基本布局属性是一致的，使用 android.support.v4.view.ViewPager 作为组件的标签。

```xml
<?xml version="1.0" encoding="utf-8"?>
<android.support.constraint.ConstraintLayout xmlns:android="http://schemas.android.com/apk/res/android"
    xmlns:app="http://schemas.android.com/apk/res-auto"
    xmlns:tools="http://schemas.android.com/tools"
    android:layout_width="match_parent"
    android:layout_height="match_parent"
    tools:context=".MainActivity">
    >
    <android.support.v4.view.ViewPager
        android:id="@+id/viewpager"
        android:layout_width="0dp"
        android:layout_height="wrap_content"
        app:layout_constraintEnd_toEndOf="parent"
        app:layout_constraintHorizontal_bias="1.0"
        app:layout_constraintStart_toStartOf="parent"
        app:layout_constraintTop_toTopOf="parent">
        <android.support.v4.view.PagerTabStrip
            android:id="@+id/pagertabstrip"
            android:layout_width="wrap_content"
            android:layout_height="wrap_content">
        </android.support.v4.view.PagerTabStrip>
    </android.support.v4.view.ViewPager>
</android.support.constraint.ConstraintLayout>
```

图 4-31　主视图中 ViewPager 的布局代码

ViewPager 由多个 Fragment 组成，每个 Fragment 的布局都一致，如图 4-32 所示的是 activity_viewpager_fragment_1.xml 的布局代码；每一个 Fragment 的布局都使用一个 ImageView，每个布局的 ImageView 展示的图片不一样。

```xml
<?xml version="1.0" encoding="utf-8"?>
<LinearLayout xmlns:android="http://schemas.android.com/apk/res/android"
    xmlns:app="http://schemas.android.com/apk/res-auto"
    android:layout_width="match_parent"
    android:layout_height="match_parent">
    <ImageView
        android:id="@+id/imageView"
        android:layout_width="match_parent"
        android:layout_height="match_parent"
        android:layout_weight="1"
        android:scaleType="centerCrop"
        android:src="@drawable/school1" />
</LinearLayout>
```

图 4-32　activity_viewpager_fragment_1.xml 的布局代码

为 Fragment1 到 Fragment7 的碎片布局新建 7 个 Fragment，继承 android.support.v4.app.Fragment 类；下面以 Fragment1 为例，代码实现如图 4-33 所示。首先重写 Fragment1 的 onCreateView()方法，然后在 onCreateView 方法中通过 LayoutInflater 的 inflate()方法将刚才定义的 activity_viewpager_fragment_1.xml 布局动态加载进来。

```
package com.example.chapter4viewpagertest;
import android.os.Bundle;
```

图 4-33　Fragment1 代码实现

```
import android.support.v4.app.Fragment;
import android.view.LayoutInflater;
import android.view.View;
import android.view.ViewGroup;
public class Fragment1 extends Fragment {
    @Override
    public View onCreateView(LayoutInflater inflater, ViewGroup container,
                             Bundle savedInstanceState) {
        return inflater.inflate(R.layout.activity_viewpager_fragment_1, container, false);
    }
}
```

图 4-33 Fragment1 代码实现（续）

自定义 FragmentPageAdapter 适配器 MyFragmentPageAdapter 的代码实现如图 4-34 所示；定义属性变量 private Context mCtx 代表主视图上下文；private ArrayList<Fragment> mFragments 代表碎片的集合数据源；使用属性变量定义构造函数，FragmentManager fm 参数代表主视图的 Fragment 管理器；实现 FragmentPageAdapter 重载方法；getItem 代表获取指定位置的 Fragment 实例数据；getCount 代表 Fragment 的数量。

```
package com.example.chapter4viewpagertest;
import android.content.Context;
import android.support.annotation.Nullable;
import android.support.v4.app.Fragment;
import android.support.v4.app.FragmentManager;
import android.support.v4.app.FragmentPagerAdapter;
import java.util.ArrayList;
import java.util.List;
public class MyFragmentPageAdapter extends FragmentPagerAdapter {
    private Context mCtx;
    private ArrayList<Fragment> mFragments;
    private List<String> mTitleList;
    public MyFragmentPageAdapter(FragmentManager fm, Context mCtx, ArrayList<Fragment> mFragments) {
        super(fm);
        this.mCtx = mCtx;
        this.mFragments = mFragments;
    }
    @Override
    public Fragment getItem(int i) {
        return mFragments.get(i);
    }
    @Override
    public int getCount() {
        return mFragments.size();
    }
}
```

图 4-34 适配器 MyFragmentPageAdapter 的代码实现

主视图 MainAcvitiy 的实现代码如图 4-35 所示：初始化 Fragment 集合数据；在 initData 方法中初始化 Fragment 集合数据 mFragments；将新建的 Fragment 依次添加到 mFragments 中；定义适配器对象 "MyFragmentPageAdapter adapter = new MyFragmentPageAdapter(getSupportFragmentManager(),this,mFragments,mTitleList);"，将适配器与 ViewPager 组件对象进行绑定 mViewpager.setAdapter(adapter)。

```
package com.example.chapter4viewpagertest;
import android.content.pm.ActivityInfo;
import android.graphics.Color;
import android.os.Bundle;
import android.support.v4.app.Fragment;
```

图 4-35 主视图 MainAcvitiy 的实现代码

```java
import android.support.v4.view.ViewPager;
import android.support.v7.app.AppCompatActivity;
import android.util.TypedValue;
import java.util.ArrayList;
import java.util.List;
public class MainActivity extends AppCompatActivity {
    private ArrayList<Fragment> mFragments;
    private ViewPager mViewpager;
    private MyFragmentPageAdapter adapter;//使用自定义FragmentPagerAdapter
        @Override
    protected void onCreate(Bundle savedInstanceState) {
        super.onCreate(savedInstanceState);
        setContentView(R.layout.activity_main);
        initView();
        initData();
        initAdapter();
    }
    private void initAdapter() {
        //使用自定义FragmentPagerAdapter
        adapter = new MyFragmentPageAdapter(getSupportFragmentManager(),this,mFragments);
        mViewpager.setAdapter(adapter);
    }
    private void initData() {
        //Fragments 初始化
        mFragments = new ArrayList<>();
        mFragments.add(new Fragment1());
        mFragments.add(new Fragment2());
        mFragments.add(new Fragment3());
        mFragments.add(new Fragment4());
        mFragments.add(new Fragment5());
        mFragments.add(new Fragment6());
        mFragments.add(new Fragment7());
    }
    private void initView() {
        mViewpager = (ViewPager) findViewById(R.id.viewpager);
    }
}
```

图 4-35 主视图 MainAcvitiy 的实现代码（续）

完成后的 ViewPager 的多页面切换的效果如图 4-36 所示。可以使用 ViewPager 组件实现了一个相册的翻页功能。

图 4-36 ViewPager 的多页面切换的效果

4.2.2 引导页与选项卡

本小节我们主要介绍 ViewPager 实现引导页；引导页一般是在用户第一次进入 App 时给用户的友好提示，包括介绍 App 的基本功能、最近更新的功能等。目前市场上的 App 的引导页大部分都采用 ViewPager 滑动的方式来实现，每一个页面采用图片方式填充。

引导页和标题栏
（慕课）

本小节实例的引导页使用 7 张图片来进行填充，并且在底部添加了一个小的操作平台，包括了 7 个单选按钮，滑动时单选按钮能动态指示当前的页面，单击单选按钮能够进入当前的页面，如图 4-37 所示。

图 4-37 引导页实例图

引导页和选项卡
（实践案例）

主视图布局代码如图 4-38 所示；底部设计了一个 RadioGroup 容器存放多个 RadioButton 按钮。单选按钮根据当前的页面位置，做隐藏和显示处理。单击单选按钮后 ViewPager 组件可以显示不同的页面。

```xml
<?xml version="1.0" encoding="utf-8"?>
<android.support.constraint.ConstraintLayout xmlns:android="http://schemas.android.com/apk/res/android"
    xmlns:app="http://schemas.android.com/apk/res-auto"
    xmlns:tools="http://schemas.android.com/tools"
    android:layout_width="match_parent"
    android:layout_height="match_parent"
    tools:context=".MainActivity"
    >
    <android.support.v4.view.ViewPager
        android:id="@+id/viewpager"
        android:layout_width="0dp"
        android:layout_height="wrap_content"
        app:layout_constraintEnd_toEndOf="parent"
        app:layout_constraintHorizontal_bias="1.0"
        app:layout_constraintStart_toStartOf="parent"
        app:layout_constraintTop_toTopOf="parent">
        <android.support.v4.view.PagerTabStrip
            android:id="@+id/pagertabstrip"
            android:layout_width="wrap_content"
            android:layout_height="wrap_content">
        </android.support.v4.view.PagerTabStrip>
    </android.support.v4.view.ViewPager>
    <android.support.constraint.Guideline
        android:id="@+id/guideline"
        android:layout_width="wrap_content"
        android:layout_height="wrap_content"
        android:orientation="horizontal"
```

图 4-38 主视图布局代码

```xml
        app:layout_constraintGuide_percent="0.9" />
<RadioGroup
        android:id="@+id/radiogroup"
        android:layout_width="wrap_content"
        android:layout_height="wrap_content"
        android:layout_marginStart="8dp"
        android:layout_marginLeft="8dp"
        android:layout_marginTop="8dp"
        android:layout_marginEnd="8dp"
        android:layout_marginRight="8dp"
        android:layout_marginBottom="8dp"
        android:orientation="horizontal"
        app:layout_constraintBottom_toBottomOf="parent"
        app:layout_constraintEnd_toEndOf="parent"
        app:layout_constraintStart_toStartOf="parent"
        app:layout_constraintTop_toTopOf="@+id/guideline">
        <RadioButton
            android:id="@+id/radioButton1"
            android:layout_width="40dp"
            android:layout_height="40dp"
            android:layout_margin="10dp"
            android:layout_weight="1"
            android:background="@drawable/tab_selector"
            android:button="@null"
            android:padding="5dp" />
        <RadioButton
            android:id="@+id/radioButton2"
            android:layout_width="40dp"
            android:layout_height="40dp"
            android:layout_margin="10dp"
            android:layout_weight="1"
            android:background="@drawable/tab_selector"
            android:button="@null"
            android:padding="5dp" />
        <RadioButton
            android:id="@+id/radioButton3"
            android:layout_width="40dp"
            android:layout_height="40dp"
            android:layout_margin="10dp"
            android:layout_weight="1"
            android:background="@drawable/tab_selector"
            android:button="@null"
            android:padding="5dp" />
        <RadioButton
            android:id="@+id/radioButton4"
            android:layout_width="40dp"
            android:layout_height="40dp"
            android:layout_margin="10dp"
            android:layout_weight="1"
            android:background="@drawable/tab_selector"
            android:button="@null"
            android:padding="5dp" />
        <RadioButton
            android:id="@+id/radioButton5"
            android:layout_width="40dp"
            android:layout_height="40dp"
            android:layout_margin="10dp"
            android:layout_weight="1"
            android:background="@drawable/tab_selector"
            android:button="@null"
            android:padding="5dp" />
        <RadioButton
            android:id="@+id/radioButton6"
            android:layout_width="40dp"
            android:layout_height="40dp"
            android:layout_margin="10dp"
```

图 4-38 主视图布局代码（续）

```xml
            android:layout_weight="1"
            android:background="@drawable/tab_selector"
            android:button="@null"
            android:padding="5dp" />
        <RadioButton
            android:id="@+id/radioButton7"
            android:layout_width="40dp"
            android:layout_height="40dp"
            android:layout_margin="10dp"
            android:layout_weight="1"
            android:background="@drawable/tab_selector"
            android:button="@null"
            android:padding="5dp" />
    </RadioGroup>
</android.support.constraint.ConstraintLayout>
```

图 4-38 主视图布局代码（续）

RadioButton 通过选择器 tab_selector.xml 实现图片转换，选择器 tab_selector.xml 的实现代码如图 4-39 所示；其中被选中的时候进行显式处理，显示为蓝色，未被选中则进行隐式处理，显示为灰色。

```xml
<?xml version="1.0" encoding="UTF-8"?>
<selector xmlns:android="http://schemas.android.com/apk/res/android">
    <item android:drawable="@drawable/tab_checked" android:state_checked="true"/>
    <item android:drawable="@drawable/tab_normal"/>
</selector>
```

图 4-39 选择器 tab_selector.xml 的实现代码

RadioButton 显式处理的背景如图 4-40 所示，背景颜色使用的是"@color/light_blue"。

```xml
<?xml version="1.0" encoding="utf-8"?>
<shape xmlns:android="http://schemas.android.com/apk/res/android"
    android:shape="oval">
    <solid android:color="@color/light_blue"/>
    <stroke android:color="@color/light_blue"/>
    <size
        android:width="30dp"
        android:height="30dp"/>
</shape>
```

图 4-40 RadioButton 显式处理的背景——tab_checked.xml 代码

RadioButton 隐式处理的背景如图 4-41 所示，背景颜色使用的是"@color/light_gray"。

```xml
<?xml version="1.0" encoding="utf-8"?>
<shape xmlns:android="http://schemas.android.com/apk/res/android"
    android:shape="oval">
    <solid android:color="@color/light_gray"/>
    <stroke android:color="@color/light_gray" android:width="1dp"/>
    <size
        android:width="30dp"
        android:height="30dp"/>
</shape>
```

图 4-41 RadioButton 隐式处理的背景——tab_normal.xml 代码

主视图 MainAcvitiy 代码实现如图 4-42 所示：主视图实现 RadioGroup 单选框被选中和 ViewPager 页面改变两个接口；实现 Activity 初始化函数；初始化 RadioButton 的 ID 集合类对象 mTabs；将所有的 RadioButton 的 ID 依次加入到 mTabs；切换的过程中根据 ViewPager 选

中的序号来确定 RadioButton 选中的 Button；根据 RadioButton 选中的 ID 来确定 ViewPager 的显示序号。

```java
package com.example.chapter4viewpagertest;
import android.content.pm.ActivityInfo;
import android.graphics.Color;
import android.os.Bundle;
import android.support.v4.app.Fragment;
import android.support.v4.view.PagerTabStrip;
import android.support.v4.view.ViewPager;
import android.support.v7.app.AppCompatActivity;
import android.util.TypedValue;
import android.widget.RadioGroup;
import java.util.ArrayList;
import java.util.List;
public class MainActivity extends AppCompatActivity implements ViewPager.OnPageChangeListener,RadioGroup.OnCheckedChangeListener {
    private ArrayList<Fragment> mFragments;
    private ViewPager mViewpager;
    private MyFragmentPageAdapter adapter;//使用自定义 FragmentPagerAdapter
    private RadioGroup mRadiogroup;
    private List<Integer> mTabs ;
    @Override
    protected void onCreate(Bundle savedInstanceState) {
        super.onCreate(savedInstanceState);
        setContentView(R.layout.activity_main);
        initView();
        initData();
        initAdapter();
    }
    private void initAdapter() {
        //使用自定义 FragmentPagerAdapter
        adapter = new MyFragmentPageAdapter(getSupportFragmentManager(),this,mFragments);
        mViewpager.setAdapter(adapter);
    }
    private void initData() {
        //Fragments 初始化
        mFragments = new ArrayList<>();
        mFragments.add(new Fragment1());
        mFragments.add(new Fragment2());
        mFragments.add(new Fragment3());
        mFragments.add(new Fragment4());
        mFragments.add(new Fragment5());
        mFragments.add(new Fragment6());
        mFragments.add(new Fragment7());
        //RadioButtonid 初始化
        mTabs = new ArrayList<>();
        mTabs.add(R.id.radioButton1);
        mTabs.add(R.id.radioButton2);
        mTabs.add(R.id.radioButton3);
        mTabs.add(R.id.radioButton4);
        mTabs.add(R.id.radioButton5);
        mTabs.add(R.id.radioButton6);
        mTabs.add(R.id.radioButton7);
        mRadiogroup.check(R.id.radioButton1);
    }

    private void initView() {
        mViewpager = (ViewPager) findViewById(R.id.viewpager);
        mViewpager.addOnPageChangeListener(this);
        mRadiogroup = (RadioGroup) findViewById(R.id.radiogroup);
        mRadiogroup.setOnCheckedChangeListener(this);
    }
    @Override
    public void onPageScrolled(int i, float v, int i1) {
```

图 4-42　主视图 MainAcvitiy 代码实现

```
}
@Override
public void onPageSelected(int i) {
    //根据 ViewPager 选中序号确定 RadioButton 的选中 Button
    mRadiogroup.check(mTabs.get(i));
}
@Override
public void onPageScrollStateChanged(int i) {

}
@Override
public void onCheckedChanged(RadioGroup group, int checkedId) {
//根据 RadioButton 选中 ID 确定 ViewPager 的显示序号
    mViewpager.setCurrentItem(mTabs.indexOf(checkedId));
}
@Override
protected void onResume() {
    if(getRequestedOrientation()!=ActivityInfo.SCREEN_ORIENTATION_LANDSCAPE){
        setRequestedOrientation(ActivityInfo.SCREEN_ORIENTATION_LANDSCAPE);
    }
    super.onResume();
}
}
```

图 4-42 主视图 MainAcvitiy 代码实现（续）

下面我们继续介绍 ViewPager 组件通过 PagerTabStrip 实现标题栏，具体效果如图 4-43 所示：滑动 ViewPager 组件，图片的上方对应地显示图片的标题。

图 4-43 标题栏具体效果

ViewPager 的布局代码如图 4-44 所示，ViewPager 的布局与普通组件基本布局属性是一致的，使用 android.support.v4.view.ViewPager 作为组件的标签；在 ViewPager 的布局中使用 PagerTabStrip 带下画线的标题栏。

```xml
<android.support.v4.view.ViewPager
    android:id="@+id/viewpager"
    android:layout_width="0dp"
    android:layout_height="wrap_content"
    app:layout_constraintEnd_toEndOf="parent"
    app:layout_constraintHorizontal_bias="1.0"
    app:layout_constraintStart_toStartOf="parent"
    app:layout_constraintTop_toTopOf="parent">
    <android.support.v4.view.PagerTabStrip
        android:id="@+id/pagertabstrip"
        android:layout_width="wrap_content"
        android:layout_height="wrap_content">
    </android.support.v4.view.PagerTabStrip>
</android.support.v4.view.ViewPager>
```

图 4-44 ViewPager 的布局代码

需要修改自定义 FragmentPageAdapter 子类，增加一个标题栏的文字集合属性；具体的实现如图 4-45 所示；定义属性变量 private Context mCtx 代表主视图上下文；private ArrayList<Fragment> mFragments 代表碎片的集合数据源；private List<String> mTitleList 代表标题栏的文字集合；使用属性变量定义构造函数，FragmentManager fm 参数代表主视图的视图管理器；实现 FragmentPageAdapter 重载方法；getItem 代表获取指定位置的 Fragment 实例数据；getCount 代表 Fragment 的数量；getPageTitle 代表获取标题栏的文字。

```java
package com.example.chapter4viewpagertest;
import android.content.Context;
import android.support.annotation.Nullable;
import android.support.v4.app.Fragment;
import android.support.v4.app.FragmentManager;
import android.support.v4.app.FragmentPagerAdapter;
import java.util.ArrayList;
import java.util.List;
public class MyFragmentPageAdapter extends FragmentPagerAdapter {
    private Context mCtx;
    private ArrayList<Fragment> mFragments;
    private List<String> mTitleList;
    public MyFragmentPageAdapter(FragmentManager fm, Context mCtx, ArrayList<Fragment> mFragments, List<String> mTitleList) {
        super(fm);
        this.mCtx = mCtx;
        this.mFragments = mFragments;
        this.mTitleList = mTitleList;
    }
    @Override
    public Fragment getItem(int i) {
        return mFragments.get(i);
    }
    @Override
    public int getCount() {
        return mFragments.size();
    }
    @Nullable
    @Override
    public CharSequence getPageTitle(int position) {
        return mTitleList.get(position);
    }
}
```

图 4-45　自定义 FragmentPageAdapter 子类 MyFragmentPageAdapter 代码

标题栏的实现代码如图 4-46 所示；initView 中设置 PagerTabStrip 对象，主要完成 PagerTabStrip 对象初始化；设置标题栏背景颜色；设置文本颜色；设置标题栏为长线；设置短线的颜色。

```java
private void initAdapter() {
    //设置选项卡属性
    mPagerTabStrip.setBackgroundColor(Color.LTGRAY);/*设置背景颜色*/
    mPagerTabStrip.setTextColor(Color.RED);/*设置文本颜色*/
    mPagerTabStrip.setTextSize(TypedValue.COMPLEX_UNIT_SP,40);/*设置文本颜色*/
    mPagerTabStrip.setDrawFullUnderline(true);//长线
    mPagerTabStrip.setTabIndicatorColor(Color.RED);/*设置短线颜色*/
    //使用自定义 FragmentPagerAdapter
    adapter = new MyFragmentPageAdapter(getSupportFragmentManager(),this,mFragments,mTitleList);
    mViewpager.setAdapter(adapter);
}
```

图 4-46　标题栏的实现代码

主视图 MainAcvitiy 中实现标题栏代码如图 4-47 所示：初始化标题栏 Title 数据；依次将

标题栏的文字加入到 List<String> mTitleList 集合数据中；自定义 FragmentPagerAdapter 对象初始化，并将 ViewPager 组件对象与适配器进行绑定。

```java
private List<String> mTitleList;
private String[] strTitles={"大门","图书馆","行政楼","天鹅湖","宿舍楼","体育场","广场"};
private void initData() {
    //title 初始化
    mTitleList= new ArrayList<>();
    for(int i=0;i<strTitles.length;i++){
        mTitleList.add(strTitles[i]);
    }
}
private void initAdapter() {
    //使用自定义 FragmentPagerAdapter
    adapter = new MyFragmentPageAdapter(getSupportFragmentManager(),this,mFragments,mTitleList);
    mViewpager.setAdapter(adapter);
}
```

图 4-47 主视图 MainAcvitiy 中实现标题栏代码

代码编写完成后运行程序，效果如图 4-48 所示，滑动时单选按钮能动态指示当前的页面，单击单选按钮能够进入当前的页面。

图 4-48 标题栏运行效果

4.2.3 单元小测

选择题：

1. ViewPager 类设置页面项的适配器的方法是（　　）。
   A. setCurrentItem()   B. setAdapter()
   C. onPageScrollStateChanged()   D. addOnPageChangeListener()
2. ViewPager 类设置当前页面的方法是（　　）。
   A. setCurrentItem()setAdapter()   B. onPageScroll
   C. StateChanged()   D. addOnPageChangeListener()
3. ViewPager 选中页面时，滑动结束时触发哪个方法？（　　）
   A. addOnPageChangeListener()   B. onPageScrollStateChanged()
   C. onPageScrolled()   D. onPageSelected()
4. ViewPager 滑动状态变化时触发哪个方法？（　　）

A. addOnPageChangeListener()　　　　　　B. onPageScrollStateChanged()
C. onPageScrolled()　　　　　　　　　　　D. onPageSelected()

5. ViewPager 滑动过程中触发哪个方法？（　　）

A. addOnPageChangeListener()　　　　　　B. onPageScrollStateChanged()
C. onPageScrolled()　　　　　　　　　　　D. onPageSelected()

6. ViewPager 主要使用下面哪个适配器？（　　）

A. PagerAdapter　　　B. ArrayAdapter　　　C. BaseAdapter　　　D. ListAdapter

7. 使用 RadioGroup 默认显示第一个选项，请补全下面的代码。（　　）

```
public class MainActivity extends AppCompatActivity implements ViewPager.OnPageChangeListener,
RadioGroup.OnCheckedChangeListener{
    private List<Integer> mTabs= new ArrayList<>();
    private void initData() {
        mTabs.add(R.id.radioButton1);
        mTabs.add(R.id.radioButton7);
        (    ?    );
    }
}
```

A. mRadiogroup.check(R.id.radioButton7);

B. mRadiogroup.set(R.id.radioButton7);

C. mRadiogroup.check(R.id.radioButton1);

D. mRadiogroup.findViewByid(R.id.radioButton7);

8. 启动引导页中，ViewPager 选中页面后实现 RadioGroup 对应的选项被选中，请补全代码。（　　）

```
public void onPageSelected(int i) {
    //根据 ViewPager 选中序号确定 RadioButton 的选中 Button
    (    ?    );
}
```

A. mRadiogroup.set(mTabs.get(i));

B. mRadiogroup.check(mTabs.get(i));

C. mRadiogroup.set(mTabs);

D. mRadiogroup.check(mTabs);

9. PageAdapter 必须实现的方法主要包括（　　）。（多选题）

A. getCount()　　　B. instantiateItem()　　　C. getView()　　　D. destroyItem()

10. FragmentPagerAdapter 必须实现的方法主要包括（　　）。（多选题）

A. getCount()　　　B. getItem()　　　C. instantiateItem()　　　D. destroyItem()

本章课后练习和程序源代码

第 4 章源代码及课后习题

5 Android 广播

Android 的广播收发机制、Android 的自定义广播和有序广播。

1. 熟练掌握 Android 的广播收发机制。
2. 熟练使用 Android 的自定义广播。
3. 熟练使用 Android 的有序广播。

我们上学时都有过这样的经历,当我们在火车站候车室中等待时,每当有某次列车开始检票或者进站上车时,就会播放广播通知来告知在候车室等待的人们该消息。

类似的工作机制其实在计算机领域也有很广泛的应用,在一个局域网的 IP 网络范围中,最大的 IP 地址是被保留作为广播地址来使用的。比如某个网络的 IP 范围是 192.168.1.1~192.168.1.255,子掩码是 255.255.2550,那么这个网络的广播地址就是 192.168.1.255。IP 地址为 192.168.1.255 的计算机可以发送广播数据包到同一网络上的所有端口,该网络中的每台主机都将会收到这条广播。

为了便于进行系统级别的消息通知,Android 引入了一套类似的广播机制,比上述两种情景要灵活并且功能更强大,本章就将对 Android 的广播进行详细的讲解。

 5.1 广播概述

5.1.1 Android 广播收发机制

Android 广播概述
(慕课)

我们常用的手机卫士就有很多模块使用了广播机制,下面看一下应用的具体实例,比如在手机卫士中进行电量的管理,骚扰电话的拦截和显示,骚扰短信的拦截和显示,如图 5-1

所示。

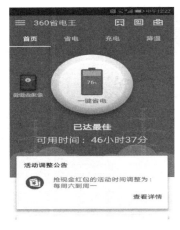  

图 5-1　手机卫士广播实例

什么是广播呢？如图 5-2 所示的是广播示意图。广播有如下的一些特征：广播是一种数据传送与交换方式；广播传送数据时存在两个固定角色，即负责发送广播的发送方和可以接收广播的接收方；广播发送方只负责按特定通道（频道）发送数据，并不考虑接收；广播接收方只有通过特定通道（频道）才能接收到数据。

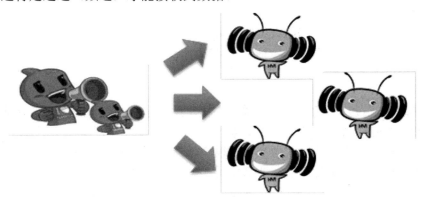

图 5-2　广播示意图

Android 中的每个应用程序都可以对自己感兴趣的广播进行注册，这样该程序就只会接收到自己所关心的广播内容。这些广播可能是来自于系统的，也可能是来自于其他应用程序的。Android 提供了一套完整的 API 允许应用程序自由地发送和接收广播。Android 广播（Broadcast）是 Android 四大核心组件之一。以广播发送者分类，广播可被分为系统广播和自定义广播。Android 广播机制是 Android 系统实现应用程序之间数据传递的一种方式，这种方式被称为跨进程间通信（IPC）。

应用程序是如何应用 Android 广播的呢？Android 会通过系统广播告知当前设备中所有应用程序系统环境或系统应用发生的变化，设备中的应用程序可以根据需要来接收这些广播，以实现相应的业务。

如图 5-3 所示的是一个手机卫士应用的例子。设备收到短信与电话后产生了 Android 来电广播和短信广播，应用程序可以拦截发送到设备中的短信与电话；还可以获取设备中电池电量的变化情况；获取设备启动、关机的信息。Android 系统是系统广播的发送方，应用程序需要编写广播接收器来接收广播。

图 5-3　一个手机卫士应用的例子——广播应用图

Android 有哪些广播呢？如图 5-4 所示的是 Android 的广播类型，通过 Android 的广播 Action 的名字就可以使用 Android 的广播，比如，我们常用的有如下几种：SMS_RECEIVED_ACTION（接收到短信时，系统发送的广播）；ACTION_PHONE_STATE_CHANGED（电话打出、接入时，系统发送的广播）；ACTION_BATTERY_LOW（设备处于低电量时，系统发送的广播）。

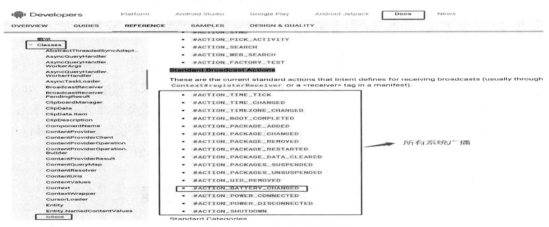

图 5-4　Android 的广播类型

Android 常用的系统广播的类型和说明如表 5-1 所示。

表 5-1　Android 常用的系统广播的类型和说明

序号	广播 Action 名称	广播的作用
1	ADD_SHORTCUT_ACTION	在系统中添加一个快捷方式时候，系统发送的广播
2	ACTION_PHONE_STATE_CHANGED	电话打出、接入时，系统发送的广播（该常量保存在 TelephonyManager 类中）
3	ACTION_NEW_OUTGOING_CALL	外拨电话时，系统发送的广播
4	SMS_RECEIVED_ACTION	接收到短信时，系统发送的广播（该常量保存在 Telephony 类中）

续表

序号	广播 Action 名称	广播的作用
5	CONFIGURATION_CHANGED_ACTION	通过设定菜单改变 Android 设置时，系统发送的广播
6	DATE_CHANGED_ACTION	日期被改变，系统发送的广播
7	BOOT_COMPLETED_ACTION	Android 系统启动时，系统发送的广播
8	BATTERY_CHANGED_ACTION	充电状态，或者电池的电量发生变化时，系统发送的广播
9	ACTION_BATTERY_LOW	设备处于低电量时，系统发送的广播
10	ACTION_BATTERY_OKAY	设备充电完成后，系统发送的广播
11	ACTION_POWER_CONNECTED	设备连接充电器后，系统发送的广播
12	ACTION_POWER_DISCONNECTED	设备断开充电器后，系统发送的广播

我们自己开发的应用程序如何接收 Android 的系统广播呢？可以使用如下的步骤。

◆ 获取常用的 Android 系统广播种类。
◆ 理解常用系统广播所使用的 Action（频道/通道）。
◆ 编写广播接收类（BroadcastReceiver）。
◆ 按相应的 Action（频道）将接收器注册到 Android 系统。
◆ 在接收器内编写对应广播数据的处理业务。

如果我们要获取 Android 设备中的电池信息，比如当前设备是否处于充电状态？当前设备使用何种充电模式？当前设备的实际电量？当前设备充电是否完成？当前设备的电池是否在低电量环境中运行？

手机电量显示
（实践案例）

Android 通过广播机制传递电池的状态信息：设备接入充电器时 Android 会在系统范围内发送设备充电广播；设备拔出充电器时 Android 会在系统范围内发送"设备停止充电"广播；设备电池电量发生变化时，Android 会在系统范围内发送"设备电池电量发生改变"的广播；每次发送广播时，广播中附带了大量的电池设备的信息。

5.1.2 实践案例——获取设备中电池的电量

下面我们以一个实例讲解如何获取设备中电池的电量，具体步骤如下：

手机充电状态显示
（实践案例）

◆ 获取 Android 系统广播的类型。
◆ 获取电池电量广播的 Action（频道）。
◆ 编写广播接收类"BatteryReceiver"接收广播。
◆ 向 Android 系统注册广播接收器"BatteryReceiver"。
◆ 从广播中获取电池信息。
◆ 应用程序退出时注销广播接收器。

Android 有很多内部组件，比如线程、服务、广播、视图、组件之间的通信有很多种方式，常用的有 Handler 机制如图 5-5 所示；线程视图一般通过使用视图 Handler 发送消息到视图的消息队列,视图从消息队列中取出消息,更新视图 UI 组件信息;这种机制使得组件之间的通信比较复杂,组件之间不能解耦,代码移植和可读性都不好。

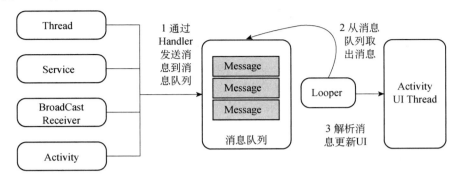

图 5-5　Android 的 Handler 机制

EventBus 是一个开源库的 Android 和 Java 库,使用发布/订阅方式,采用的是松耦合模式。EventBus 使中央通信能够用简单的几行代码来解耦类,可以简化代码,消除依赖关系,加速应用程序的开发。组件之间使用 EventBus,接收数据的组件将自己注册到 EventBus,发送数据的组件将数据通过 POST 发送到总线,总线将数据发送到订阅者,如图 5-6 所示。

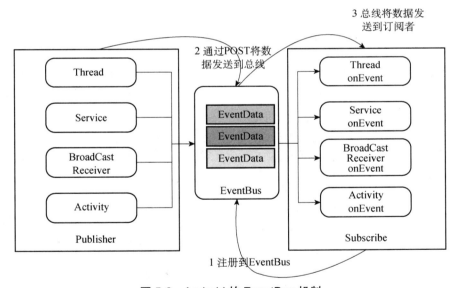

图 5-6　Android 的 EventBus 机制

EventBus 总线的作用主要有如下几个特点:
◆ 简化组件之间的通信。
◆ 解耦事件的发送者和接收者。
◆ 使用 UI(例如,activity、fragment)和后台线程避免复杂易出错的依赖和生命周期问题。
◆ 快速,专门优化代码,具有高性能。

如何使用 EventBus 呢?在 Build.gradle 文件的 dependencies 上添加插件依赖 implementation 'org.greenrobot:eventbus:3.1.1'，如图 5-7 所示。

```
dependencies {
    implementation 'org.greenrobot:eventbus:3.1.1'
}
```

图 5-7　EventBus 的 dependencies 添加代码

EventBus 初始化，将 Activity 注册到 EventBus 上 "EventBus.getDefault().register(this);"，如图 5-8 所示。

```
private void initData() {
    EventBus.getDefault().register(this);
}
```

图 5-8　EventBus 的初始化代码

EventBus 的消息发送：新建 EventData 事件对象 EventData mEventData=new EventData()；设置事件类型 mEventData.setEventCode(1)；设置数据 mEventData.setiBatterydata(result)；发送数据 EventBus.getDefault().post(mEventData)，如图 5-9 所示。

```
EventData mEventData=new EventData();
mEventData.setEventCode(1);
mEventData.setiBatterydata(result);
EventBus.getDefault().post(mEventData);
```

图 5-9　EventBus 的消息发送代码

如何处理 EventBus 的消息呢？如图 5-10 所示，使用@Subscribe 表示的是 EventBus 的消息处理；重新实现 public void onEventMainThread(EventData event)方法；使用 event.getEventCode() 获取消息类型；使用 event.getiBatterydata()取出数据。

```
@Subscribe
public void onEventMainThread(EventData event) {
    if (event == null) {
        return;
    }
    if (event.getEventCode() == 1) {
        mTvBattery.setText("电池电量状态:" +
                event.getiBatterydata() + "%");
    }
}
```

图 5-10　EventBus 的消息处理代码

EventBus 如何注销呢？如图 5-11 所示，重写视图的 onDestroy 方法；使用 EventBus.getDefault().unregister(this)注销 EventBus 的订阅。

```
protected void onDestroy() {
    super.onDestroy();
    EventBus.getDefault().unregister(this);
}
```

图 5-11　EventBus 的注销代码

下面以一个电量广播实例讲解如何实现 Android 系统广播的接收，如图 5-12 所示。

图 5-12 电量广播实例

首先我们编写广播接收类"BatteryReceiver"接收广播；继承 BroadcastReceiver 并重写 onReceive 方法，代码如图 5-13 所示。

```java
package cn.edu.sziit.chapter5_batteray;
import android.content.BroadcastReceiver;
import android.content.Context;
import android.content.Intent;
import android.os.BatteryManager;
import android.os.Bundle;
import org.greenrobot.eventbus.EventBus;
public class BatteryReceiver extends BroadcastReceiver {
    @Override
    public void onReceive(Context context, Intent intent) {
        String strAction=intent.getAction();
        if (Intent.ACTION_BATTERY_CHANGED.equals(strAction)) {
            Bundle bundle = intent.getExtras();
            int level = bundle.getInt(BatteryManager.EXTRA_LEVEL, -1);
            int scale = bundle.getInt(BatteryManager.EXTRA_SCALE, -1);
            int result = (int) ((float) level / (float) scale * 100);
            EventData mEventData=new EventData();
            mEventData.setEventCode(1);
            mEventData.setiBatterydata(result);
            EventBus.getDefault().post(mEventData);
        }
    }
}
```

图 5-13 广播接收类 BatteryReceiver 实现代码

首先使用 intent.getAction() 取出意图的名称；判断意图是否为电量改变的广播 ACTION_BATTERY_CHANGED；使用 intent.getExtras() 方法取出广播数据；根据 BatteryManager.EXTRA_LEVEL 取出电量绝对值；根据 BatteryManager.EXTRA_SCALE 取出

电量最大值；换算出电量的百分比 int result = (int) ((float) level / (float) scale * 100)；将电量值使用 EventBus 发送到 EventBus 总线上。

在 Activity 中注册广播接收器"BatteryReceiver"，其代码如图 5-14 所示。定义广播接收器 private BatteryReceiver mBatteryReceiver；将 Activity 视图活动注册到 EventBus；初始化广播接收器对象；新建电量变化广播过滤器；设置广播接收器接收电量广播。

```
private BatteryReceiver mBatteryReceiver ;定义广播接收器对象；
    private void initData() {
    EventBus.getDefault().register(this);//将 Activity 获得注册到 EventBus
    mBatteryReceiver = new BatteryReceiver();//初始化广播接收器对象
    IntentFilter filter = new
    IntentFilter(Intent.ACTION_BATTERY_CHANGED); //新建电量变化广播过滤器
    super.registerReceiver(mBatteryReceiver, filter); //设置广播接收器接收电量广播
}
```

图 5-14　注册广播接收器 BatteryReceiver 代码

新建 EventData 类，BatteryReceiver 与 MainActivity 通过 EventData 在 EventBus 总线上传递数据，如图 5-15 所示，新建 eventCode 变量代表数据类型；新建的 iBatterydata 变量代表电量值。

```
public class EventData {
    //eventCode=1 代表电量事件；eventCode=2 代表充电事件；
    private int eventCode;//数据类型
    private int iBatterydata;//电量值
    private boolean bCharge;
}
```

图 5-15　新建 EventData 类代码

Activity 中使用 EventBus 获取信息；电池电量信息均被保存在 EventData 对象中，其代码如图 5-16 所示：重写 Event 的消息处理方法 onEventMainThread；使用 event.getEventCode() 判断是否是电量变化广播消息；使用 event.getiBatterydata() 取出电量的数据，并显示到组件上。

```
public void onEventMainThread(EventData event) {
    if (event.getEventCode() == 1) {
        mProgressBarBatteray.setProgress(event.getiBatterydata());
        mTvBattery.setText("电池电量状态:" + event.getiBatterydata() + "%");
    }
}
```

图 5-16　EventBus 获取信息代码

应用程序退出后注销 EventBus 和广播接收器，其代码如图 5-17 所示：视图中重写 onDestroy() 方法；使用 super.unregisterReceiver(mBatteryReceiver) 注销广播接收器；使用 EventBus.getDefault(). unregister(this) 方法注销 EventBus 接收。

```
protected void onDestroy() {
    super.onDestroy();
    super.unregisterReceiver(mBatteryReceiver);
    EventBus.getDefault().unregister(this);
}
```

图 5-17　注销 EventBus 和广播接收器代码

代码编写完成后可以在手机上运行程序，程序运行的效果如图 5-18 所示，可以看到手机电量的实时显示情况。

图 5-18　程序运行的效果——手机电量的实时显示

5.1.3　单元小测

判断题：

每一个无序广播只能有一个广播接收器接收。（　　）

A. 是　　　　　　　　　　　　　　B. 否

选择题：

1. 下列方法中，用于发送一条无序广播的是（　　）。

A. startBroadcastReceiver()　　　　　　B. sendOrderedBroadcast()

C. sendBroadcast()　　　　　　　　　　C. sendReceiver()

2. 下面实现动态注册广播，请补全代码。（　　）

```
AnotherBroadCastReceiver mAnotherBroadCastReceiver ; //定义广播接收器
private void initData() {
mAnotherBroadCastReceiver = new AnotherBroadCastReceiver(); //初始化广播接收器对象
    //新建自定义广播过滤器
IntentFilter filter = new IntentFilter(AnotherBroadCastReceiver.BROADCAST_TYPE1);
(   ?   )
}"
```

A. super.register(mAnotherBroadCastReceiver, filter)

B. super.registerBroadCast(mAnotherBroadCastReceiver, filter)

C. super.registerReceiver(mAnotherBroadCastReceiver, filter)
D. super.registerBroadCastReceiver(mAnotherBroadCastReceiver, filter)

3. 下面方法实现发送无序广播，请补全代码。（　　）

```
private void sendBroadcast() {
    Intent intent=new Intent(MyReceiver.BROADCAST_TYPE1); //创建广播意图
    (　?　)
}
```

A. startBroadcastReceiver(intent)　　　　B. sendOrderedBroadcast(intent)
C. sendBroadcast(intent)　　　　　　　　D. sendReceiver(intent)

5.2　广播收发机制

5.2.1　知识点讲解——广播收发机制

Android 广播工作原理
（慕课）

本节我们介绍广播接收器的工作流程，广播接收器的处理流程如图 5-19 所示；广播接收器的生命周期取决于 onReceive 方法。广播接收器一旦处理完业务，也就是 onReceive 执行完毕，广播接收器对象将会处于失活状态（No-longer Active），Android 系统有权在内存资源紧张的时候回收广播接收器。不要在 onReceive 中编写耗时操作，onReceive 方法运行于应用的主线程中，一旦处理耗时操作将会导致 ANR 错误。注销广播接收器后广播接收器将不会再收到任何广播数据。

图 5-19　广播接收器的处理流程

Android 中的广播注册流程如图 5-20 所示。Android 系统中有一个广播接收器注册列表 RegisteredReceivers；使用 super.registerReceiver 可以注册广播到广播接收器，比如 CustomerActivity 中注册了一个短信广播；NetworkService 中注册了一个网络状态的广播。

所有的广播接收器都会被注册到 ActivityManagerService 中，registerReceiver 方法通过动态方式注册接收器。PackageMangerService 扫描所有 Android 应用的 AndroidManifest.xml，将配置文件中的静态配置的接收器注册入 ActivityManagerService 中。注册完接收器后，为接收器设置的 Action（频道）同时会与广播接收器捆绑，所有的广播接收器都被注册在 RegisteredReceivers 列表中。

当一个广播发送到 ActivityManagerService 中，满足广播 Action 的接收器将会获取广播的对应数据，也就是接收器的 onReceive 方法将会被回调调用。

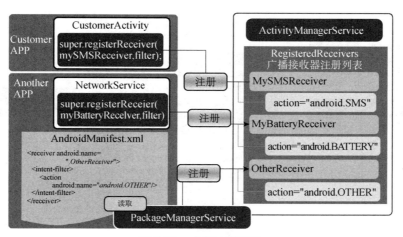

图 5-20 广播注册流程

广播接收器的注册分为动态注册和静态注册，Android7.0 后只提供动态注册。动态注册广播接收器的代码如图 5-21 所示。使用 context. registerReceiver 方法注册广播接收器。注册广播接收器时需要提供广播接收器和对象广播接收器 Action（频道）。

```
private NetworkStatusReceiver mNetworkStatusReceiver;
    private void initData() {
        mNetworkStatusReceiver = new NetworkStatusReceiver();
        IntentFilter filter4 = new
            IntentFilter("android.net.conn.CONNECTIVITY_CHANGE");
        super.registerReceiver(mNetworkStatusReceiver, filter4);
}
}
```

图 5-21 动态注册广播接收器的代码

静态注册广播接收器的代码如图 5-22 所示；广播接收器被注册在 AndroidManifest.xml 配置文件中。注册广播接收器时需要提供广播接收器类全名和广播接收器 Action（频道）。

```
<application
<receiver
        android:name=". PowerConnectReceiver"
        android:enabled="true"
        android:exported="true">
        <intent-filter>
            <action android:name="android.net.conn.CONNECTIVITY_CHANGE" />
        </intent-filter>
    </receiver>
</application>
```

图 5-22 静态注册广播接收器的代码

广播接收器接收广播的原理如图5-23所示；所有的广播均会被发送到ActivityManagerService。收到广播后，ActivityManagerService 从 RegisteredReceivers 列表中寻找满足当前广播 Action 的 Receiver。将满足 Action 的 Receiver 保存入队列中准备发送广播。依次发送广播给队列中的 Receiver。Receiver 中保存了应用的 PID（进程 ID）。Receiver 收到 ActivityManagerService 传递过来的数据。Receiver 的 onReceive 可以接收到本次数据传递。

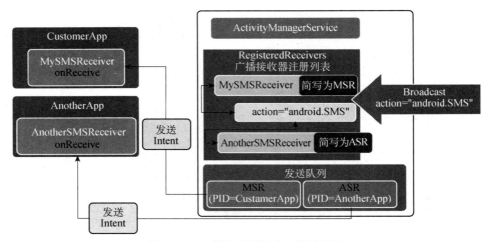

图 5-23　广播接收器接收广播的原理

5.2.2　实践案例——显示网格状态

下面以一个网络状态显示的实例来学习广播的发送和接收流程；网络不能使用的时候我们希望手机应用程序能够有所提示，在如图 5-24 所示的例子中，微信应用程序在网络状态不能连接的时候会提示"网络连接不可用"。

图 5-24　微信的网络状态实例

网络状态是如何实现的呢？如图 5-25 所示，我们以一个显示网络状态的实例来讲解如何获取系统的网络广播，当网络是 WiFi 的时候，显示 WiFi 连接；当网络是 4G 信号的时候，显示 4G 网络。

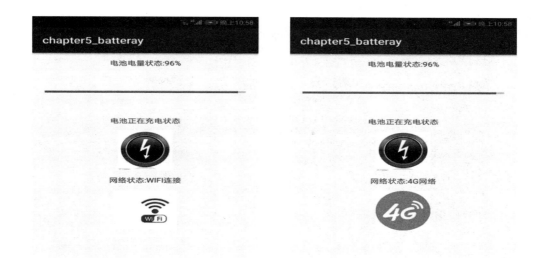

图 5-25 网络状态显示实例

下面我们介绍如何实现网络状态广播的接收：编写广播接收类"NetworkStatusReceiver"接收广播，继承 BroadcastReceiver 并重写 onReceive 方法，如图 5-26 所示。首先使用 intent.getAction() 取出意图的名称；判断意图是否为网络状态改变的广播 CONNECTIVITY_CHANGE；使用自定义的 getAPNType 方法判断网络类型；将网络状态的编号和数据使用 EventBus 发送到 EventBus 总线上。

```java
package cn.edu.sziit.chapter5_batteray;
import android.content.BroadcastReceiver;
import android.content.Context;
import android.content.Intent;
import android.net.ConnectivityManager;
import android.net.NetworkInfo;
import android.telephony.TelephonyManager;
import org.greenrobot.eventbus.EventBus;

public class NetworkStatusReceiver extends BroadcastReceiver {
    @Override
    public void onReceive(Context context, Intent intent) {
        // an Intent broadcast.

        if("android.net.conn.CONNECTIVITY_CHANGE".equals
(intent.getAction())) {
            int iType=getAPNType(context);
            EventData mEventData=new EventData();
            mEventData.setEventCode(3);
            mEventData.setiBatterydata(iType);
            EventBus.getDefault().post(mEventData);
        }
    }
    public int getAPNType(Context context) {
        int netType = 0;//无网络
        ConnectivityManager connMgr = (ConnectivityManager) context
                .getSystemService(Context.CONNECTIVITY_SERVICE);
        NetworkInfo networkInfo = connMgr.getActiveNetworkInfo();
        if (networkInfo == null) {
```

图 5-26 NetworkStatusReceiver 实现代码

```
        return netType;
    }
    int nType = networkInfo.getType();
    if (nType == ConnectivityManager.TYPE_WIFI) {
        netType = 1;// wifi
    } else if (nType == ConnectivityManager.TYPE_MOBILE) {
        int nSubType = networkInfo.getSubtype();
        TelephonyManager mTelephony = (TelephonyManager) context
                .getSystemService(Context.TELEPHONY_SERVICE);
        if (nSubType == TelephonyManager.NETWORK_TYPE_UMTS
                && !mTelephony.isNetworkRoaming()) {
            netType = 2;// 3G
        } else {
            netType = 3;// 4G
        }
    }
    return netType;
}
```

图 5-26　NetworkStatusReceiver 实现代码（续）

新建 getAPNType 方法用于分析并获取网络状态。首先使用 getSystemService 获得系统的网络管理器；使用 getActiveNetworkInfo 获取系统的网络详细信息；使用 getType() 获取网络类型；如果网络的类型为移动信号，使用 networkInfo.getSubtype() 获取子网络类型，最后返回网络类型。

在 Activity 中注册广播接收器 "NetworkStatusReceiver"，如图 5-27 所示；定义广播接收器；将 Activity 活动注册到 EventBus；初始化广播接收器对象；新建网络状态广播过滤器；设置广播接收器接收网络状态广播。

```
private NetworkStatusReceiver mNetworkStatusReceiver; //定义广播接收器
private void initData() {
    mNetworkStatusReceiver = new NetworkStatusReceiver();//初始化广播接收器对象
//新建网络状态广播过滤器
    IntentFilter filter4 = new
    IntentFilter("android.net.conn.CONNECTIVITY_CHANGE");
//设置广播接收器接收网络状态广播
    super.registerReceiver(mNetworkStatusReceiver, filter4); }
```

图 5-27　NetworkStatusReceiver 实现代码

NetworkStatusReceiver 与 MainActivity 通过 EventBus 传递数据，新建 EventData 类，如图 5-28 所示；新建 eventCode 变量代表数据类型；新建 iBatterydata 变量代表电量值。

```
public class EventData {
    //eventCode=1 代表电量事件；eventCode=2 代表充电事件；
    private int eventCode;//数据类型
    private int iBatterydata;//电量值
    private boolean bCharge;
}
```

图 5-28　EventData 类代码

Activity 中使用 EventBus 获取信息；电池电量信息均被保存在 EventData 对象中，重写 Event 的消息处理方法 onEventMainThread，如图 5-29 所示，使用 event.getEventCode() 方法判断是否是网络状态类型广播消息，使用 event.getiBatterydata() 取出网络状态类型的数据，并显示到组件上。

```
@Subscribe
public void onEventMainThread(EventData event) {
    if (event == null) {
        return;
    }
    if (event.getEventCode() == 3) {
        if (event.getiBatterydata()==0) {
            mTvNetwork.setText("网络不通");
            mImgNetwork.setImageResource(R.drawable.disconnent);
        }else if (event.getiBatterydata()==1) {
            mTvNetwork.setText("网络状态:WIFI 连接");
            mImgNetwork.setImageResource(R.drawable.wifi);
        }else if (event.getiBatterydata()==2) {
            mTvNetwork.setText("网络状态:4G 网络");
            mImgNetwork.setImageResource(R.drawable.fourg);
        }else if(event.getiBatterydata()==3) {
            mTvNetwork.setText("网络状态:4G 网络");
            mImgNetwork.setImageResource(R.drawable.fourg);
        }else
        {
            mTvNetwork.setText("网络状态:未知网络");
            mImgNetwork.setImageResource(R.drawable.wifi);
        }
    }
}
```

图 5-29　Event 的消息处理方法 onEventMainThread 代码

应用程序退出，注销 EventBus 和广播接收器，代码如图 5-30 所示。视图中重写 onDestroy() 方法；使用 super.unregisterReceiver(mNetwork StatusReceiver)注销广播接收器；使用 EventBus.getDefault().unregister(this)方法注销 EventBus 接收器。

```
protected void onDestroy() {
    super.onDestroy();
    super.unregisterReceiver(mNetworkStatusReceiver);
    EventBus.getDefault().unregister(this);
}
```

图 5-30　注销 EventBus 和广播接收器代码

5.2.3　单元小测

判断题：

1. 每一个广播只能有一个广播接收器接收。（　　）
A. 是　　　　　　　　　　　　　　B. 否
2. 广播接收者注册后必须在程序中手动关闭。（　　）
A. 是　　　　　　　　　　　　　　B. 否

选择题：

1. 继承 BroadcastReceiver 需要重写什么方法？（　　）
A. onUpdate()　　　B. onCreate()　　　C. onStart()　　　D. onReceiver()
2. 清单文件中，注册广播使用的节点是（　　）。
A. <activity>　　　B. <broadcast>　　　C. <receiver>　　　D. <broadcastreceiver>

3. 关于 BroadcastReceiver，下列说法不正确的是（　　）。
A. 用于接收系统或者程序中的广播事件
B. 一个广播事件只能被一个广播接收者接收
C. 对于有序广播，系统会根据接收者声明的优先级按照顺序依次接收
D. 接收者的优先级在 android：priority 中声明，数值越大优先级越高
4. 设置网络状态监听的广播，请补全代码。（　　）

```
private NetworkStatusReceiver mNetworkStatusReceiver; //定义广播接收器
    private void initData() {
    mNetworkStatusReceiver = new NetworkStatusReceiver();//初始化广播接收器对象
    IntentFilter filter4 = new
    IntentFilter("android.net.conn.CONNECTIVITY_CHANGE");
    (　？　);
}
```

A. super.unRegisterReceiver(mNetworkStatusReceiver, filter4);
B. super.registerReceiver(mNetworkStatusReceiver, filter4);
C. super.register(mNetworkStatusReceiver, filter4);
D. super.unRegister(mNetworkStatusReceiver, filter4);

5.3　自定义广播

5.3.1　知识点讲解——自定义广播

自定义广播（慕课）

上一节已经学习了通过广播接收器来接收系统广播，当系统级别的广播事件不能满足实际需求时，还可以自定义广播，接下来我们就来学习一下如何在应用程序中发送自定义广播。图 5-31 所示的是一个发送自定义广播的实例，本应用程序可以收到广播消息并显示；另外一个应用程序也可以收到这条广播消息，自定义的广播消息也可以跨进程接收。

图 5-31　发送自定义广播的实例

Android 系统发送的广播被称为 Android 系统广播；Android 应用程序发送的广播被称为自定义广播。发送自定义广播主要使用 sendBroadcast 方法，其代码如图 5-32 所示。

```
public    void sendBroadcast (Intent intent, String receiverPermission)
public    void sendBroadcast (Intent intent)
```

图 5-32   sendBroadcast 方法代码

sendBroadcast 方法的参数含义如下：第一个参数 Intent intent 为广播意图。意图中包含了广播接收 Action（频道）和广播数据。第二个参数 String receiverPermission 是本广播的访问权限。

5.3.2   实践案例——实现自定义广播

下面我们介绍一下自定义广播实例的实现过程：编写广播接收类 "MyReceiver" 接收广播；继承 BroadcastReceiver 并重写 onReceive 方法，如图 5-33 所示；首先定义广播名称和收到广播消息后提示。

```
public class MyReceiver extends BroadcastReceiver {
//定义广播名称
    public static String BROADCAST_TYPE1="cn.edu.sziit.broadcast.MY_BRODCAST";
    @Override
    public void onReceive(Context context, Intent intent) {
        Toast.makeText(context,"Receiver in
          MyReceiver ",Toast.LENGTH_SHORT).show(); //收到广播消息后提示
    }
}
```

图 5-33   广播接收类 MyReceiver 代码

在 Activity 中注册广播接收器 "MyReceiver"，如图 5-34 所示；定义广播接收器；初始化广播接收器对象；新建自定义广播过滤器；设置广播接收器接收自定义广播。

```
private MyReceiver myReceiver;//定义广播接收器
private void initData() {
    myReceiver = new MyReceiver();//初始化广播接收器对象
//新建自定义广播过滤器
    IntentFilter filter = new IntentFilter(MyReceiver.BROADCAST_TYPE1);
    super.registerReceiver(myReceiver, filter); //设置广播接收器接收自定义广播
}
```

图 5-34   Activity 中注册广播接收器代码

在 Activity 中创建按钮，单击后发送广播，新建自定义 sendBroadcast 方法，其代码如图 5-35 所示；创建广播意图，广播接收 Action（频道）为 BROADCAST_TYPE1；使用 sendBroadcast 方法发送广播。

```
private void sendBroadcast() {
    Intent intent=new Intent(MyReceiver.BROADCAST_TYPE1); //创建广播意图
    sendBroadcast(intent);//发送广播
}
```

图 5-35   自定义 sendBroadcast 方法代码

下面我们介绍一下跨进程接收广播的实现过程：新建一个广播应用程序，编写广播接收类"AnotherBroadCastReceiver"继承 BroadcastReceiver 并重写 onReceive 方法，如图 5-36 所示；首先定义广播名称和收到广播消息后提示。

```java
public class AnotherBroadCastReceiver extends BroadcastReceiver {
//首先定义广播名称
    public static String BROADCAST_TYPE1="cn.edu.sziit.broadcast.MY_BRODCAST";
public void onReceive(Context context, Intent intent) {
//收到广播消息后提示
        Toast.makeText(context,getClass().toString(),Toast.LENGTH_SHORT).show();
    }
}
```

图 5-36　AnotherBroadCastReceiver 广播接收类代码

在 Activity 中注册广播接收器，其代码如图 5-37 所示。定义广播接收器；初始化广播接收器对象；新建自定义广播过滤器；设置广播接收器接收自定义广播。

```java
AnotherBroadCastReceiver mAnotherBroadCastReceiver ; //定义广播接收器
private void initData() {
//初始化广播接收器对象
    mAnotherBroadCastReceiver = new AnotherBroadCastReceiver();
//新建自定义广播过滤器
    IntentFilter filter = new IntentFilter(AnotherBroadCastReceiver.BROADCAST_TYPE1);
//设置广播接收器接收自定义广播
    super.registerReceiver(mAnotherBroadCastReceiver, filter);
}
```

图 5-37　Activity 中注册广播接收器代码

程序编写完成后的运行实例如图 5-38 所示，在应用程序中发送自定义广播，本应用程序可以收到广播消息并显示；另外一个应用程序也可以收到这条广播消息，这个实例实现了自定义的广播消息的跨进程接收。

图 5-38　自定义广播运行实例

5.3.3 单元小测

判断题：

每一个无序广播只能有一个广播接收器接收。（　　）
A. 是　　　　　　　　　　　　　　B. 否

选择题：

1. 下列方法中，用于发送一条无序广播的是（　　）。
A. startBroadcastReceiver()　　　　　　B. sendOrderedBroadcast()
C. sendBroadcast()　　　　　　　　　　D. sendReceiver()
2. 下面实现动态注册广播，请补全代码。（　　）

```
AnotherBroadCastReceiver mAnotherBroadCastReceiver ; //定义广播接收器
private void initData() {
mAnotherBroadCastReceiver = new AnotherBroadCastReceiver(); //初始化广播接收器对象
    //新建自定义广播过滤器
IntentFilter filter = new IntentFilter(AnotherBroadCastReceiver.BROADCAST_TYPE1);
（　？　）
}
```

A. super.register(mAnotherBroadCastReceiver, filter)
B. super.registerBroadCast(mAnotherBroadCastReceiver, filter)
C. super.registerReceiver(mAnotherBroadCastReceiver, filter)
D. super.registerBroadCastReceiver(mAnotherBroadCastReceiver, filter)
3. 下面方法实现发送无序广播，请补全代码。（　　）

```
private void sendBroadcast() {
    Intent intent=new Intent(MyReceiver.BROADCAST_TYPE1); //创建广播意图
    （　？　）
}
```

A. startBroadcastReceiver(intent)　　　　B sendOrderedBroadcast(intent)
C. sendBroadcast(intent)　　　　　　　　D sendReceiver(intent)

 ## 5.4 有序广播

5.4.1 知识点讲解——有序广播

有序广播（慕课）

上一节已经学习了发送自定义广播。广播分为有序广播和无序广播两种；普通的广播一般是无序广播，用 sendBroadcast()方法发送。普通广播是完全异步的，逻辑上可以在同一时刻被所有匹配的接收者接收到，消息传递效率高，缺点是接收者不能将处理结果传递给下一个接收者，也无法终止广播传播。有序广播，用 sendOrderedBroadcast()方法发送，即从优先级别最高的广播接收器开始接收，接收完了如果没有丢弃，就传给下一个次高优先级的广播接收器进行处理，依次类推，直到最后。如果多个应用程序设置的优先级别相同，则谁先注册的广播，谁就可以优先接收到广播。

有序广播的工作流程如图 5-39 所示。

◆ 广播发送方发送的广播会按特定次序依次传播给多个接收方。

◆ 广播接收方的接收次序由 android：priority 属性控制。

◆ android：priority 的取值范围在-1000 到 1000 之间，相同优先级的接收器接收到广播的顺序为随机。

◆ 有序广播允许接收方修改广播数据，或者取消下级接收器的广播接收权利。

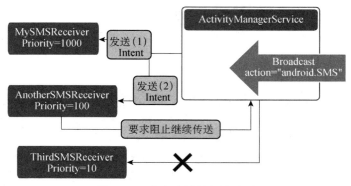

图 5-39　有序广播的工作流程

有序广播的工作流程图中有 3 个接收短消息广播接收器，优先级分别为 1000、100 和 10；系统广播产生时，优先级为 1000 的广播先收到广播消息；然后优先级为 100 的广播接收器收到消息，并且取消了广播消息；优先级为 10 的广播接收器接收不到消息。

有序广播有哪些作用呢？具体的作用就是实现骚扰短信的拦截，如图 5-40 所示；Android 的短信广播就是一个典型的有序广播；Android 系统在收到短信后会立刻发送一个广播，广播 Action 值为 android.provider.Telephony.SMS_RECEIVED。

图 5-40　有序广播的作用

有序广播最大的特点是，所有被注册的广播接收器必须严格按照 android：priority 数值在 ActivityManagerService 的接收队列中按大小排列。手机卫士等应用程序可以设置接收短消息的最大优先级，抢先接收短信后，如果是黑名单或者骚扰短信，就阻止 Android 短信收件箱获取短信，就能实现拦截垃圾短信的目的。

有序广播的工作特点如下：
◆ 广播数据按顺序依次被传递给广播接收器。
◆ 任何应用都可以接收有序广播。
◆ 广播接收器通过优先级依次获取广播。
◆ 优先级高的广播接收器可以决定后续接收器是否能获取广播。
◆ 优先级高的广播接收器可以修改广播数据，并将修改结果往下级传递。

登录广播——
记住密码
（实践案例）

5.4.2 实践案例——实现有序广播

上一节我们实现了一个无序广播的实例，发送无序广播，应用程序自己可以收到广播消息并显示；另外一个应用程序也可以收到这条广播消息；这一节我们来实现有序广播，发送自定义广播，只有应用程序自己可以收到，收到后关闭广播消息，阻止其他的应用程序收到。

登录广播——
强制退出
（实践案例）

下面我们介绍一下有序广播实例的实现过程。编写广播接收类"MyReceiver"接收广播；继承 BroadcastReceiver 并重写 onReceive 方法，其代码如图 5-41 所示；首先定义广播名称；收到广播消息后提示；收到广播后丢弃，阻止其他应用程序收到。

```java
public class MyReceiver extends BroadcastReceiver {
//定义广播名称
    public static String BROADCAST_TYPE1="cn.edu.sziit.broadcast.MY_BRODCAST";
    @Override
    public void onReceive(Context context, Intent intent) {
        Toast.makeText(context,"Receiver in
          MyReceiver",Toast.LENGTH_SHORT).show(); //收到广播消息后提示
abortBroadcast();// 收到广播后丢弃
    }
}
```

图 5-41 广播接收类 MyReceiver 代码

在 Activity 中注册广播接收器"MyReceiver"，如图 5-42 所示；定义广播接收器；初始化广播接收器对象；新建自定义广播过滤器；设置广播接收器的优先级；设置广播接收器接收自定义广播。

```java
private MyReceiver myReceiver;//定义广播接收器
private void initData() {
    myReceiver = new MyReceiver();//初始化广播接收器对象
//新建自定义广播过滤器
    IntentFilter filter = new IntentFilter(MyReceiver.BROADCAST_TYPE1);
filter.setPriority(100);//设置广播接收器的优先级
    super.registerReceiver(myReceiver, filter); //设置广播接收器接收自定义广播
}
```

图 5-42 Activity 中注册广播接收器代码

在 Activity 中创建按钮，单击后发送广播，新建自定义 sendBroadcast 方法，其代码如图

5-43 所示；创建广播意图，广播接收 Action（频道）为 BROADCAST_TYPE1；使用 sendOrdered Broadcast 方法发送有序广播。

```
private void sendBroadcast() {
    Intent intent=new Intent(MyReceiver.BROADCAST_TYPE1);//定义广播意图
    sendOrderedBroadcast(intent,null); //发送有序广播
}
```

图 5-43　sendBroadcast 方法代码

应用程序编写完成后在手机中运行程序，在应用程序中发送有序广播，应用程序自己可以收到广播消息并显示；应用程序收到后阻止其他的应用程序收到广播消息，如图 5-44 所示。

图 5-44　有序广播运行实例

5.4.3　单元小测

判断题：

每一个有序广播只能有一个广播接收器接收。（　　）
A. 是　　　　　　　　　　　　　　　B. 否

选择题：

1. 下列方法中，用于发送一条有序广播的是（　　）。
A. startBroadcastReceiver()　　　　　　　B. sendOrderedBroadcast()
C. sendBroadcast()　　　　　　　　　　　D. sendReceiver()
2. 关于 BroadcastReceiver，下面说法不正确的是（　　）。
A. 用于接收系统或者程序中的广播事件
B. 一个广播事件可以被多个广播接收者接收
C. 对于有序广播，系统会根据接收者声明的优先级按照顺序依次接收
D. 接收者的优先级在 android:priority 中声明，数值越小优先级越高

3. 下面实现动态注册广播，请补全代码。（　　）

```
private MyReceiver myReceiver;//定义广播接收器
private void initData() {
    myReceiver = new MyReceiver();//初始化广播接收器对象
    IntentFilter filter = new IntentFilter(MyReceiver.BROADCAST_TYPE1);//新建自定义广播过滤器
    filter.setPriority(100);//设置广播接收器的优先级
     （   ?   ）
}
```

A. super.registerReceiver(myReceiver, filter)

B. super.registerBroadCastReceiver(myReceiver, filter) \

C. super.register(myReceiver, filter)

D. super.registerBroadCast(myReceiver, filter)

4. 下面方法实现发送有序广播，请补全代码。（　　）

```
private void sendBroadcast() {
Intent intent=new Intent(MyReceiver.BROADCAST_TYPE1); //创建广播意图        （   ?   ）
}
```

A. startBroadcastReceiver(intent)　　　　B. sendOrderedBroadcast(intent)

C. sendBroadcast(intent)　　　　　　　　D. sendReceiver(intent)

本章课后练习和程序源代码

第 5 章源代码及课后习题

6 Android 系统服务

Android 的系统服务、Android 的自定义服务、Android 的多线程机制。

1. 熟练掌握和使用 Android 的系统服务。
2. 熟练使用 Android 的自定义服务。
3. 熟练使用 Android 的多线程。

Android 的 Service 是 Context 的子类，Service 是四大组件之一，用来在后台处理一些比较耗时的操作或者去执行某些需要长期运行的任务；服务的运行不依赖用户界面，即使服务被切换到后台，或者用户切换到另外一个应用程序，服务仍然能够保持独立运行。

服务并不运行于一个独立的进程，而是依赖于创建服务所在的应用程序进程；当应用程序进程被终止的时候，所依赖于该进程的服务也会停止运行。

服务不会自动开启线程，所有的服务默认为运行到主线程，我们需要在服务的内部自己新建子线程去执行具体的耗时的操作和任务，否则也会堵塞子线程，本章我们将对 Android 系统服务的使用进行详细的讲解。

 ## 6.1 系统服务概述

6.1.1 Android 的服务组件

系统服务概述（慕课）

本节我们介绍 Android 的服务组件，在讲解 Android 的服务之前，我们先来了解一下什么是服务，常用操作系统 Windows 服务有如下特征（图 6-1 所示的是 Windows 后台启动的服务）。

- 服务与运行在 Windows 操作系统中的应用程序都是进程。
- 服务是一个特殊的进程，特点在于没有图形用户界面（GUI）。Windows 的后台服务没有图形用户界面。
- 服务在操作系统的后台（Background）为用户默默提供各种业务。

图 6-1　Windows 后台启动的服务

Windows 后台服务主要可以实现如下的功能。

① 监控服务：
- Windows Firewall（Windows 防火墙）。
- Windows Update（自动更新 Windows 系统）。
- Shell Hardware Detection（自动播放服务）。
- XLServicePlatform（迅雷下载监控服务）。

② 耗时计算服务：
- Windows Search（Windows 全盘搜索服务）。
- Apache Tomcat（Tomcat Web 容器服务）。

③ 访问硬件服务：
- Windows Audio（系统音频服务）。
- NVIDIA Display Driver Service（视频显示硬件驱动服务）。

Android Service 作为 Android 四大组件之一，在每一个应用程序中都扮演着非常重要的角色。它主要用于在后台处理一些耗时的逻辑运算，或者去执行某些需要长期运行的任务。必要的时候我们甚至可以在程序退出的情况下，让 Service 在后台继续保持运行状态。

Android Service 与 Activity 的相同点如下：
- Android Service 和 Android Activity 都是一个组件。
- 都能够通过研发人员的编码为用户提供各种业务功能。

Android Service 与 Activity 的区别如下：
- Activity 主要用来提供图形用户界面（GUI）供用户与应用程序进行交互，并允许用户通过 GUI 向 Android 应用发送各种业务命令。
- Service 只担任"默默无闻"的业务计算工作。
- 实际项目研发过程中 Service 往往用来计算与处理用户的各种复杂和耗时的指令。
- Activity 则提供美观大方的 GUI 界面接收用户的指令，并将后台服务计算完毕（例如：通过服务计算完毕）的结果显示在界面上。

Android 从服务提供方来说可以分为系统服务和自定义服务，如图 6-2 所示。
- Android 系统服务：Android 系统运行后就能使用，比如图中后台运行的 Service1 和 Service2。
- Android 自定义服务：由研发人员根据不同的需要自行开发的服务。比如图中自定义的 Java Class1 和 Java Class2；这些服务往往与活动组件等其他组件保存在同一个应用程序中。

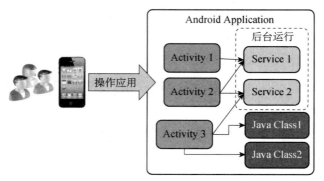

图 6-2　Android 服务分类——服务提供方

Android 服务从服务访问级别上可以分为本地服务和全局服务，如图 6-3 所示。

（1）Android 本地服务
- 只有服务的"自身程序"（宿主进程）中的应用，可以访问该服务。
- 图中的 Local Service 是一个组件，只能创建在 Application1 中，本地服务只有在该应用程序中的其他组件可以访问。
- 其他（不包含该 Service）的 Android 应用程序不得访问，比如图中的 Application2 不能访问。

（2）Android 全局服务
- 创建的服务组件可以被"宿主应用"（宿主进程）和 Android 中的其他应用访问。
- Android 全局服务是最常见的服务，比如图中的 Remote Service 可以被 Application2 访问。
- Android 的系统服务全部为全局服务，可供任何应用访问。

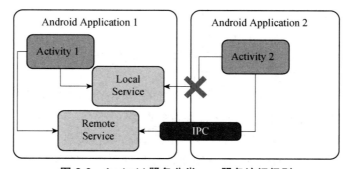

图 6-3　Android 服务分类——服务访问级别

Android 系统服务的创建过程如图 6-4 所示。

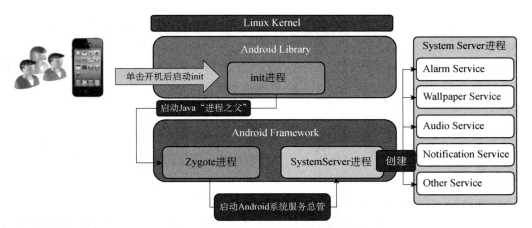

图 6-4 Android 系统服务的创建过程

① Android 系统启动时开启 init 进程，init 负责启动 Zygote 进程和 Library 中的系统级资源与进程。

② Zygote 进程负责初始化 Application Framework 中的系统级资源和进程，系统服务主要包括：
- PackageManagerService 应用程序管理服务。
- SystemServer 系统服务。
- WindowManagerService 窗口管理服务。
- ActiviyManagerService 活动管理服务。

③ System Server 进程中的核心组件是 com.android.server.SystemServer；System Server 组件负责创建、保存、管理所有系统服务。

Android 服务主要应用于如下场景。

（1）监控系统状态
- Alarm Service（提醒服务）：监控系统时间。
- TelephonyManagerService（来电服务）：监听通话语音。
- Storage Manager（磁盘服务）：磁盘空间不足 10% 时给用户警告。

（2）耗时计算服务
- Search Service（搜索服务）：搜索系统资源。
- Connectivity Service（连接服务）：实时返回各种网络连接的服务。

（3）访问硬件服务
- Power Service（电源服务）：通过访问电池获取电源信息。
- Audio Service（音频服务）：通过访问多媒体设备提供声音。
- Sensor Service（传感器服务）：识别各种传感器并注册入系统。

（4）实现跨应用程序间的通信（IPC / RPC）

服务提供的数据可以供多个应用程序中的组件访问。

6.1.2 单元小测

判断题：

1. Android 可以在子线程中直接更新主线程 UI 组件。（　　）
A. 是　　　　　　　　　　　　　　B. 否

2. Activity 以绑定的方式开启 Service 后，Activity 与 Service 在不同的线程中。（　　）
　A. 是　　　　　　　　　　　　　　B. 否
3. Activity 以绑定的方式开启 IntentService 服务后，Activity 与 IntentService 在不同的线程中。（　　）
　A. 是　　　　　　　　　　　　　　B. 否
4. AsyncTask 可以处理异步任务。（　　）
　A. 是　　　　　　　　　　　　　　B. 否

选择题：

1. IntentService 的启动方式为（　　）。
　A. bindService()　　　B. startService()　　　C. bindIntenService()　　　D. startIntenService()
2. AsyncTask 的启动方式为（　　）。
　A. startAsyncTask()　　B. startService()　　　C. execute()　　　D. startIntenService()

6.2　访问系统服务

6.2.1　Android 的系统服务组件

访问系统服务（慕课）

本节我们介绍如何访问 Android 的系统服务组件，在上一节中我们已经了解了系统服务的流程如下：

◆ Android 系统在启动时会启动 System Server 进程。
◆ System Server 进程按照特定顺序启动各个系统服务。
◆ 访问系统进程的数据需要访问 System Server 中的服务。

系统服务 System Server 的访问流程如图 6-5 所示。

◆ System Server 是一个独立进程，直接访问 System Server 需要通过大量 IPC 技术实现，操作比较复杂。
◆ Android 系统为了降低系统服务的编程难度，研发了大量的 Manager 组件。这些独立的管理组件可以简化访问系统服务的难度。
◆ Manager 组件负责 Activity 与 System Server 中系统服务进行的交互。
◆ 通过 Manager 组件提供的方法可以方便访问系统服务。

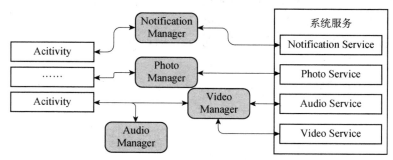

图 6-5　系统服务 System Server 的访问流程

当某个应用程序希望向用户发出一些提示信息，而该应用程序又不在前台运行时，就可以借助通知来实现，通知的使用实例如图 6-6 所示；发出一条通知后，手机最上方的状态栏中会显示一个通知的图标，下拉状态栏后可以看到通知的详细内容。

NotificationManager 是 Android 中的一个系统服务组件，可将通知对象发送到系统 ActionBar 上。该组件的主要作用是通知用户事件的发生，方法主要有如下 3 种：在状态栏中出现一个图标；设备 LED 灯闪烁；播放音乐或者是振动。

Android 提供 NotificationManager 组件，负责与 system_server 进程进行交互；获取系统通知服务管理 NotificationManager 的代码如图 6-7 所示：getSystemService 获取所有系统服务的接口，NOTIFICATION_SERVICE 代表获取通知服务。

图 6-6　通知的使用实例

```
// 获取 NotificationManager 的系统通知服务管理类
NotificationManager manager = (NotificationManager) getSystemService(NOTIFICATION_SERVICE);
```

图 6-7　获取系统通知服务管理 NotificationManager

NotificationManager 常用方法的说明如下。
◆ cancel(int id)：取消之前显示的一个通知。
◆ cancelAll()：取消之前显示的所有通知。
◆ notify(int id, Notification notification)：将通知发送到状态栏上，其中 id 代表识别符。

6.2.2　实践案例——实现通知服务

下面我们看一条通知的实例，如图 6-8 所示，需要使用 NotificationCompat 来构造通知，新建 Buider 构造方法；使用 setContentTitle 设置通知标题；使用 setContentText 设置通知内容；使用 setWhen 设置通知时间；使用 setSmallIcon 设置通知的小图标；使用 setLargeIcon 设置大图标。

通知服务
（实践案例）

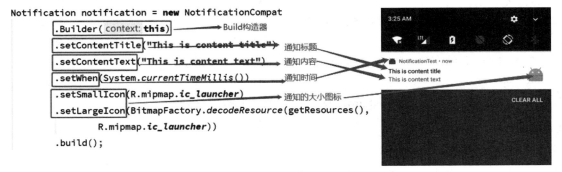

图 6-8　实现通知实例——获取系统通知服务管理 NotificationManager

完成后单击通知没有反应,这是什么原因呢?因为没有实现通知单击后的启动页面,下面我们使用 PendingIntent 来实现这个功能,代码如图 6-9 所示。新建跳转的 Activity 对象;新建延迟 PendingIntent 对象;通知中设置 PendingIntent;在 NotificationActivity 中取消已阅读通知。

```java
private void sendNotice() {
    Intent intent = new Intent(this, NotificationActivity.class);
    PendingIntent pi = PendingIntent.getActivity(this, 0, intent, 0);
    NotificationManager manager = (NotificationManager) getSystemService(NOTIFICATION_SERVICE);
    Notification notification = new NotificationCompat
            .Builder(this)
            .setContentIntent(pi)
            .setContentTitle("This is content title")
            .setContentText("This is content text")
            .setWhen(System.currentTimeMillis())
            .setSmallIcon(R.mipmap.ic_launcher)
            .setLargeIcon(BitmapFactory.decodeResource(getResources(),R.mipmap.ic_launcher))
            .build();
    manager.notify(1, notification);
}
```

图 6-9 获取系统通知服务管理 NotificationManager 代码

通知还有一些高级功能,使用"(.setSound(Uri.fromFile(new File("/system/media/audio/ringtones/Luna.ogg"))"来设置声音。

使用"setVibrate(new long[]{0, 1000, 1000, 1000});<uses-permission android:name="android.permission.VIBRATE" />"来设置震动。

使用".setLights(Color.GREEN, 1000, 1000)"来设置 LED。

使用".setDefaults(NotificationCompat.DEFAULT_ALL)"来设置默认效果。

6.2.3 单元小测

判断题:

1. 服务的界面可以设置得很美观。( )
   A. 是　　　　　　　　　　　　　　　　B. 否
2. 服务中可以处理长时间耗时的操作。( )
   A. 是　　　　　　　　　　　　　　　　B. 否
3. Android 系统在启动时会启动 System Server 进程。( )
   A. 是　　　　　　　　　　　　　　　　B. 否
4. NotificationManager 是 Android 中的一个系统服务组件。( )
   A. 是　　　　　　　　　　　　　　　　B. 否

选择题:

1. 与 System Server 中系统服务交互的管理组件有哪些?( )
   A. Notification Manager　　　　　　　　B. Photo Manager

C. Video Manager   D. Audio Manager

2. 在系统中定义并获取通知服务的代码是哪一个？（　　）

A. NotificationManager manager = (NotificationManager) getService(NOTIFICATION_SERVICE);
B. NotificationManager manager = (NotificationManager) getService(NOTIFICATION_SERVICE_Manager);
C. NotificationManager manager = (NotificationManager) getSystemService(NOTIFICATION_SERVICE_Manager);
D. NotificationManager manager = (NotificationManager) getSystemService(NOTIFICATION_SERVICE);

6.3　自定义服务

自定义服务（慕课）

6.3.1　知识点讲解——自定义服务

本节我们介绍如何使用自定义服务。在上一讲中我们详细介绍了系统服务，Android 系统为了降低系统服务的编程难度，研发了大量 Manager 组件。这些独立的管理组件可以简化访问系统服务的难度。

自定义服务跟 Activity 的级别差不多，但不能自己运行只能后台运行，研发人员根据不同的需要自行开发的服务，可以和其他组件进行交互，如图 6-10 所示。Service 可以在很多场合的应用中使用，比如播放多媒体的时候，用户启动了其他 Activity，这个时候程序要在后台继续播放，还有比如检测 SD 卡上文件的变化，再或者在后台记录你地理信息位置的改变等，服务总是藏在后台的，往往与活动组件等其他组件保存在同一个应用程序中。

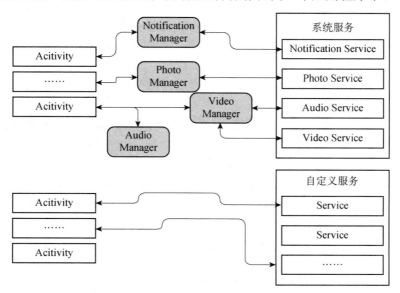

图 6-10　自定义服务

Service 生命周期没有 Activity 生命周期那么复杂，如图 6-11 所示；Service 的启动有两种方式：context.startService() 和 context.bindService()。

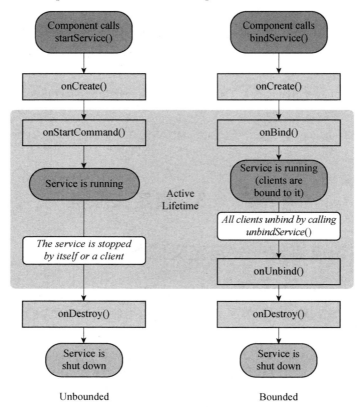

图 6-11　Service 生命周期

startService()：同一应用在任意地方调用 startService()方法都能启动 Service，然后系统会自动调用 onCreate()、onStartCommand()，这样启动的 Service 会一直在后台运行，直到 stopService()方法被调用。

如果一个 Service 已经被启动，其他代码再想调用 startService()方法，则不会执行 onCreate()，但会重新执行一次 onStartCommand()。

6.3.2　实践案例——调用 Service 生命周期函数

下面我们以一个实例来学习一下 Service 生命周期的函数调用关系：新建一个继承于 Service 的类，代码如图 6-12 所示；依次进行下列重载实现方法 onCreate()、onStartCommand()，onDestroy()，onBind()；依次在上述的重载函数中使用 Log 标签标记。

```
public class MyService extends Service {
    public void onCreate() {
        super.onCreate();
        Log.d("MyService", "onCreate executed");
    }
    public int onStartCommand(Intent intent, int flags, int startId) {
```

图 6-12　继承于 Service 的类代码

启动服务
（实践案例）

绑定服务
（实践案例）

后台服务
（实践案例）

```
        Log.d("MyService", "onStartCommand executed");
        return super.onStartCommand(intent, flags, startId);
    }
    public void onDestroy() {
        super.onDestroy();
        Log.d("MyService", "onDestroy executed");
    }
    public IBinder onBind(Intent intent) {
        Log.d("MyService", "onBind executed");
        return null;}
}
```

图 6-12　继承于 Service 的类代码（续）

实现服务的开启和关闭：使用 startService(startIntent)开启服务；使用 stopService(stopIntent)关闭服务，其代码如图 6-13 所示。

```
Intent startIntent = new Intent(this, MyService.class);
startService(startIntent); // 启动服务
Intent stopIntent = new Intent(this, MyService.class);
stopService(stopIntent); // 停止服务
```

图 6-13　服务的开启和关闭代码

在 XML 配置信息文件中注册 MyService 服务，其代码如图 6-14 所示。

```
<service
    android:name=".MyService"
    android:enabled="true"
    android:exported="true" />
<service android:name=".MyIntentService" />
```

图 6-14　注册 MyService 服务代码

如图 6-15 所示的是使用 startService 开启服务的生命周期展示效果图；第一次单击"开始服务"按钮启动服务，系统会自动调用 onCreate()、onStartCommand()；第二次单击"开始服务"按钮启动服务，则不会执行 onCreate()，但会重新执行一次 onStartCommand()；单击"关闭服务"按钮，系统会调用 onDestroy()方法。

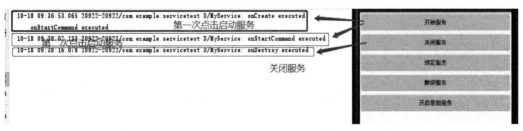

图 6-15　使用 startService 开启服务的生命周期展示效果图

下面我们介绍绑定服务 bindService()。使用 bindService()方法启动时，系统会自动调用 onCreate()、onBind()，Service 会和客户端绑定起来，客户端停止则 Service 也会停止。

绑定服务中 onBind()将返回给客户端一个 IBind 接口实例，IBind 允许客户端回调服务的方法，比如得到 Service 的实例、运行状态或其他操作。这个时候调用者（例如，Activity）会和 Service 绑定在一起，Activity 退出了，Service 就会调用 onUnbind->onDestroy 相应退出。

如图 6-16 所示的是一个下载应用程序的具体实例，主 Activity 视图绑定 Service。

① 服务启动了一个下载功能，下载功能模块包括开始下载和获取下载进度。
② Activity 如果用 startService 启动服务，Activity 不能获取下载的进度。
③ Activity 使用 bindService 绑定服务，Activity 通过 onBind 获取了下载功能的实例对象；可以使用服务中的下载功能接口获取下载进度。

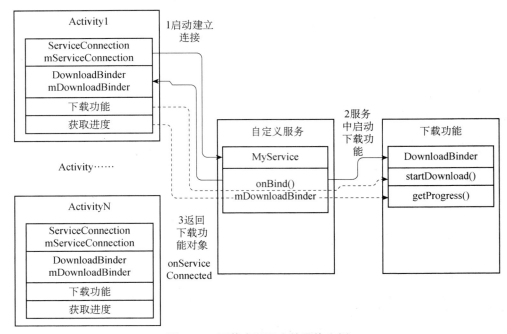

图 6-16 下载应用程序的具体实例

新建一个继承于 Binder 的 DownloadBinder 类，依次实现 startDownload 和 getProgress 这两个接口，其代码如图 6-17 所示。

```java
public class DownloadBinder extends Binder {
    public void startDownload() {
        Log.d("DownloadBinder", "startDownload executed");
    }
    public int getProgress() {
        Log.d("DownloadBinder", "getProgress executed");
        return 0;
    }
}
```

图 6-17 DownloadBinder 类代码

MyService 类中新建一个 DownLoadBinder 对象，onBind 接口中返回 DownLoadBinder 对象，其代码如图 6-18 所示。

```java
private DownloadBinder mDownloadBinder = new DownloadBinder();
@Override
public IBinder onBind(Intent intent) {
    return mDownloadBinder;
}
```

图 6-18 MyService 类代码

Activity 定义 ServiceConnection 类和 DownLoadBinder 对象并初始化，其代码如图 6-19 所示，ServiceConnection 对象的初始化必须实现 onServiceDisconnected 回调方法，绑定服务成功后，服务通过这个回调方法，将服务 MyService 中的 DownLoadBinder 对象通过 IBinder service 对象传回，Activity 中的 mDownloadBinder 对象就可以调用服务中的下载功能接口。

```
private ServiceConnection mServiceConnection;
private DownloadBinder mDownloadBinder;
mServiceConnection = new ServiceConnection() {
    public void onServiceDisconnected(ComponentName name) {
    }
    public void onServiceConnected(ComponentName name, IBinder service) {
        mDownloadBinder = (DownloadBinder) service;
    }
};
```

图 6-19　ServiceConnection 类和 DownLoadBinder 类对象初始化代码

Activity 中绑定 MyService 服务，代码如图 6-20 所示。新建一个服务意图；通过 bindService 将 Activity 与服务绑定。

```
Intent bindIntent = new Intent(this, MyService.class);//服务意图
bindService(bindIntent, mServiceConnection, BIND_AUTO_CREATE); // 绑定服务
unbindService(mServiceConnection); // 解绑服务
```

图 6-20　Activity 中绑定 MyService 服务代码

Activity 中调用 Myservice 服务中的下载和获取进度功能，代码如图 6-21 所示。

```
mDownloadBinder.startDownload();//开始下载
mDownloadBinder.getProgress();//获取下载进度
```

图 6-21　Activity 中调用 Myservice 服务代码

最后我们看一下程序运行的效果，如图 6-22 所示，单击"绑定服务"按钮后，系统会自动调用 onCreate()、onBind()两个方法；单击"开始下载"和"获取下载进度"按钮，调用了下载功能的 startDownload 和 getProgress 两个接口。单击"解绑服务"按钮，系统会调用 onDestroy 方法。

图 6-22　Activity 中调用 Myservice 服务运行效果

根据 Log 的日志，我们可以看到视图和服务都处于同一个线程，如图 6-23 所示；服务不会自动开启线程，所有的服务默认为运行到主线程。

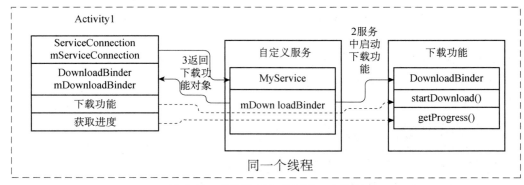

图 6-23 视图和服务处于同一个线程

6.3.3 单元小测

判断题：

1. 以绑定的方式开启服务后，服务与调用者没有关系。（　　）
   A. 是　　　　　　　　　　　　　　　B. 否
2. 以绑定的方式开启服务后，当界面不可见时服务就会关闭。（　　）
   A. 是　　　　　　　　　　　　　　　B. 否
3. 服务不需要在配置信息文件中注册。（　　）
   A. 是　　　　　　　　　　　　　　　B. 否

选择题：

1. 如果服务已启动，使用 startService 再次启动服务时，执行的生命周期方法有（　　）。
   A. onCreate()　　　　　　　　　　　B. onResume on
   C. StartCommand()　　　　　　　　D. onDestroy()
2. 配置信息文件中，注册服务使用的节点是（　　）。
   A. activity　　　　　　　　　　　　B. service
   C. receiver　　　　　　　　　　　　D. broadcastreceiver

多选题：

1. 使用 startService 启动服务时，执行的生命周期方法有（　　）。
   A. onCreate()　　　　　　　　　　　B. onResume()
   C. onStartCommand()　　　　　　　D. onDestroy()
2. 使用 bindService 启动服务时，执行的生命周期方法有（　　）。
   A. onCreate()　　　　　　　　　　　B. onBind()
   C. onStart()　　　　　　　　　　　　D. onDestroy()
3. 使用 bindService 启动服务时，下列说法正确的是（　　）。
   A. 调用者与服务没有关系　　　　　　B. 调用者关闭后服务关闭
   C　必须实现 ServiceConnection　　　D. 使用 stopService 关闭服务
4. 下列方式中，不属于服务的生命周期的是（　　）。

A. onResume B. onStop()
C. onStart() D. onDestroy()

 ## 6.4 多线程

6.4.1 知识点讲解——多线程

多线程（慕课）

本节我们介绍如何使用多线程；手机程序应用启动后系统会创建一个主线程（Main Thread）；这个主线程负责向 UI 组件分发事件，主线程中的应用和 Android 的 UI 组件发生交互，所以主线程也称为 UI 线程。

在 UI 线程中如果做一些比较耗时的工作比如访问网络或者数据库查询，都会阻塞 UI 线程，导致事件停止分发，用户体验感觉程序比较卡，没有反应。因此，创建主线程有两个原则：不要阻塞 UI 线程；不要在 UI 线程外访问 UI 组件。

下面我们看一个 UI 线程做一些比较耗时的工作比如访问网络的实例，如图 6-24 所示。比如常见的 QQ 登录程序，输入用户名和密码登录后，程序要访问服务器网络，由于需要从服务器下载好友列表，有时候网络速度也许不流畅，服务器的访问可能会比较耗时，如果在主线程中访问网络，会阻塞 UI 线程，程序会比较卡，用户体验就不好。

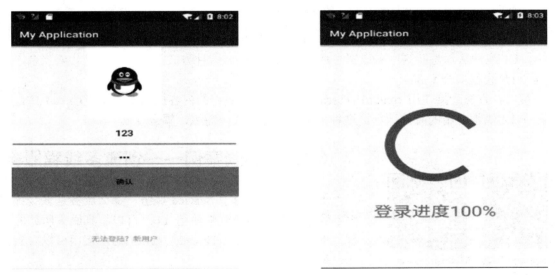

图 6-24 访问网络的实例

怎么解决应用中的耗时问题呢？如图 6-25 所示，Android 提供了多线程的模式，用新的线程 myRun 访问网络，但是它违反了主线程第二条原则：从非 UI 线程访问 UI 组件，会导致未定义和不能预料的行为。为了解决这个问题，Android 又提供了一些方法，实现其他线程访问 UI 线程。

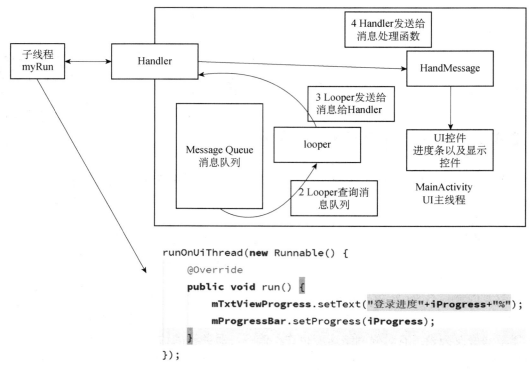

图 6-25 多线程模式

第一种方法是使用 Handler 机制。获取子线程的 Handler，将数据放入 UI 线程的消息队列中，UI 线程从消息队列中获取消息后更新 UI 组件，这种方式的不足之处是代码冗余，子线程和 UI 线程耦合度高。

第二种方法直接使用 runOnUiThread 机制。将子线程的内容放入 runOnUiThread 中，子线程在 UI 线程空闲的时候运行，这种机制的好处是代码简练，耦合度低。

6.4.2 实践案例——创建多线程服务

上一节我们学习了 Service 服务，那么服务是多线程的吗？我们在服务的函数中使用 Log 可以看到服务所处的线程，通过 Thread.currentThread(). getId()这个接口可以看到日志，如图 6-26 所示。

Log.*d*("**MainActivity**", "**Thread id is** " + Thread.*currentThread*(). getId());

图 6-26 Android 中获取线程号

主视图 MainActivity 与开启的服务 Service 的线程号是一致的；Service 虽然在后台运行，但是服务与视图本质上处于同一个线程，如图 6-27 所示；视图、服务及服务启动的下载功能本质上都在同一个线程。

MyService 下载
（实践案例）

MyIntentService
（实践案例）

AsyncTask 实现
对话框下载
（实践案例）

下载精灵
ServiceBinder
AsynTask
（实践案例）

图 6-27　服务与视图处于同一个线程

如果在启动服务中增加了一些耗时操作，如图 6-28 所示，在服务的 onStartCommand 中使用 Thread.sleep(5000)来阻塞线程 5 秒钟。

```
try {
    Log.d(strTag, "onStartCommand 线程 ID:" + Thread.currentThread().getId());
    Log.d(strTag, "文件下载......");
    Thread.sleep(5000);
    Log.d(strTag, "文件下载完成。");
}catch (InterruptedException e){
    e.printStackTrace();
}
```

图 6-28　onStartCommand 增加耗时操作代码

程序的运行结果如图 6-29 所示；启动服务后，服务与主线程是同一个线程，服务中耗时 5 秒钟，UI 主线程也阻塞了 5 秒钟，此时 UI 控件毫无反应，程序卡住了，因此，服务中不能完成耗时的操作。

图 6-29　耗时操作运行结果

Android 为了解决服务中不能完成耗时的操作，引入了 IntentService 支持异步任务的处理，如图 6-30 所示，IntentService 是继承并处理异步请求的一个类，在 IntentService 内有一个工作线程 onHandleIntent 来处理耗时操作，启动 IntentService 的方式和启动传统的 Service 一样，同时，当任务执行完后，IntentService 会自动停止，而不需要我们手动去控制或 stopSelf()。另外，可以启动 IntentService 多次，而每一个耗时操作会以工作队列的方式在 IntentService 的 onHandleIntent 回调方法中执行。

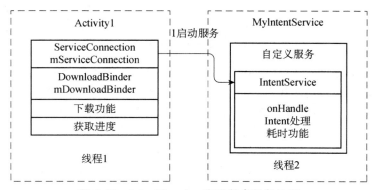

图 6-30　IntentService 实现异步任务处理

新建一个 MyIntentService 继承于 IntentService 类，如图 6-31 所示，在 onHandleIntent 方法中使用 Thread.sleep(5000)来阻塞线程 5 秒钟。

```java
public class MyIntentService extends IntentService {
    public MyIntentService() {
        super("MyIntentService"); // 调用父类的有参构造函数}
    protected void onHandleIntent(Intent intent) {
        try {
            Log.d("MyIntentService", "onHandleIntent ID:" +
Thread.currentThread().getId());
            Log.d("MyIntentService", "文件下载。。。。。");
            Thread.sleep(5000);
            Log.d("MyIntentService", "文件下载完成。");
        }catch (InterruptedException e){
            e.printStackTrace();}
    }
    public void onDestroy() {
        super.onDestroy();
        Log.d("MyIntentService", "onDestroy executed");
    }
}
```

图 6-31　MyIntentService 类代码

在主视图 Activitiy 中启动 MyIntentService，如图 6-32 所示；使用 Thread.currentThread().getId()获取主视图和服务的线程号，使用 startService 启动 MyIntentService。

```java
Log.d("MyIntentService", "Thread id is " + Thread.currentThread().getId());
Intent intentService = new Intent(this, MyIntentService.class);
startService(intentService);
```

图 6-32　Activitiy 中启动 MyIntentService 代码

Activitiy 启动 MyIntentService 的效果如图 6-33 所示，服务启动后，Activity 与服务不在同一个线程，onHandleIntent 中的 Thread.sleep(5000)耗时操作并没有阻塞主线程，程序没有出现卡顿的现象，服务完成后自动注销。

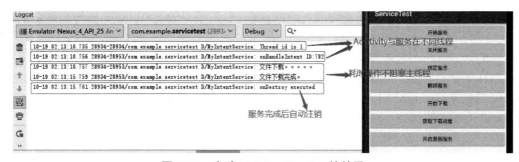

图 6-33　启动 MyIntentService 的效果

下面我们介绍 AsyncTask。AsyncTask 是一种轻量级的异步任务类，它可以在线程池中执行后台任务，然后把执行的进度和最终的结果传递给主线程并在主线程中更新 UI。

从实现上来说，AsyncTask 封装了 Thread 和 Handler，通过 AsyncTask 可以更加方便地执行后台任务及在主线程中访问 UI，但是不适合执行特别耗时的后台任务，对于特别耗时的任务来说，需要使用线程池。

AsyncTask 是一个抽象的泛型类，它提供了 Params、Progerss 和 Result 这三个泛型参数，如图 6-34 所示；其中 Params 表示参数的类型，Progress 表示后台任务执行进度的类型，而

Result 则表示后台任务的返回结果的类型,如果不需要传递具体的参数,那么这三个泛型参数可以使用 void 来代替。

```
public abstract class AsyncTask<Params, Progress, Result>
```

图 6-34　AsyncTask 定义

AsyncTask 提供了 4 个核心方法,如图 6-35 所示。
- ◆ onPreExecute():后台任务开始执行时调用,用于进行一些界面上的初始化操作。
- ◆ doInBackground(Params…):在子线程中运行,处理所有的耗时任务。
- ◆ onProgressUpdate(Progress…):在 onPreExecute 中调用了 publishProgress(Progress…)方法后被调用,参数由后台任务中传递;可以对 UI 进行操作,利用参数中的数值就可以对界面元素进行相应的更新。
- ◆ onPostExecute(Result):后台任务执行完并通过 return 语句返回时被调用。返回的数据会作为参数传递到此方法中,可以利用返回的数据来进行一些 UI 操作,比如说提醒任务执行的结果,以及关闭进度条对话框等。

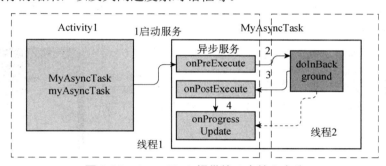

图 6-35　AsyncTask 提供的 4 个核心方法

下面我们看一下 AsyncTask 的具体实现过程,新建 MyAsyncTask 继承 AsyncTask 类,如图 6-36 所示;AsyncTask 的三个参数分别为<String,Integer,String>。

```
public class MyAsyncTask extends AsyncTask<String,Integer,String> {
    private String strTAG="MyAsyncTask";
    protected void onPreExecute() {
        Log.i(strTAG, "onPreExecute");//加上 Log 标签
    }
    protected String doInBackground(String... strings) {
        Log.i(strTAG, "doInBackground in: " + strings[0]);//启动 MyAsyncTask 传入参数
        publishProgress(1);//提交之后,会执行 onProcessUpdate 方法
        Log.i(strTAG, "doInBackground out"); //加上 Log 标签
        return "over";
    }
    protected void onCancelled() {
        Log.i(strTAG, "onCancelled"); //加上 Log 标签
    }
    protected void onPostExecute(String args3) {
        Log.i(strTAG, "onPostExecute: " + args3); //args3 为 doInBackground 返回值
    }
    protected void onProgressUpdate(Integer... args2) {
        Log.i(strTAG, "onProgressUpdate: "+ args2[0]); // args2[0] publishProgress 返回值
    }
}
```

图 6-36　AsyncTask 的实现代码

实现 AsyncTask 的 onPreExecute、doInBackground、onPostExecute、onProgressUpdate 的方法并加上标签。

doInBackground 中的参数 strings 为启动 MyAsyncTask 传入的参数；使用 publishProgress(1) 后，会执行 onProcessUpdate 方法。

onPostExecute 参数 args3 为 doInBackground 返回值。

onProgressUpdate 参数 args2 为 publishProgress 返回值。

AsyncTask 启动方法如图 6-37 所示，新建 AsyncTask 任务对象后，使用 AsyncTask 的 execute 方法启动任务，并传入参数 AsyncTask test1。

```
MyAsyncTask myAsyncTask=new MyAsyncTask();//新建 AsyncTask 任务对象
myAsyncTask.execute("AsyncTask test1");//启动 AsyncTask 任务
```

图 6-37　AsyncTask 的启动方法

程序编写完成后在手机或者模拟器中运行，AsyncTask 实例效果如图 6-38 所示。MainActivity 中启动 AsyncTask；后台任务开始执行时调用 onPreExecute；在子线程中运行 doInBackground；子线程中运行完 publishProgress 函数后，onProgressUpdate 被调用；子线程结束后调用 onPostExecute。

图 6-38　AsyncTask 实例效果

6.4.3　单元小测

判断题：

1. Android 可以在子线程中直接更新主线程 UI 组件。（　　）
   A. 是　　　　　　　　　　　　　B. 否
2. Activity 以绑定的方式开启 Service 后，Activity 与 Service 在不同的线程。（　　）
   A. 是　　　　　　　　　　　　　B. 否
3. Activity 以绑定的方式开启 IntentService 服务后，Activity 与 IntentService 在不同的线程。（　　）
   A. 是　　　　　　　　　　　　　B. 否
4. AsyncTask 可以处理异步任务。（　　）
   A. 是　　　　　　　　　　　　　B. 否

选择题：

1. IntentService 的启动方式为（　　）。
   A. bindService()　　　　　　　　B. startService()

C. bindIntenService() D. startIntenService()

2. AsyncTask 的启动方式为（　　）。

A. startAsyncTask() B. startService()

C. execute() D. startIntenService()

本章课后练习和程序源代码

第 6 章源代码及课后习题

7 Android 内容提供者

Android 的运行权限、Android 的 URL 与 URI、Android 的 ContentProvider。

1. 熟练应用 Android 的运行权限。
2. 熟练使用 Android 的 ContentProvider。
3. 熟练使用 Android 的通讯录。

7.1 Android 运行权限

7.1.1 知识点讲解——运行权限

Android 运行权限
（慕课）

Android 开发常常遇到的一个问题就是在 Android App 安装的过程中，会向用户请求一大堆权限，不同意的话不会让你安装，不知不觉中，也许有些敏感权限就这样被授予了，为了解决这个问题，Android 6.0 后推出了运行权限，敏感权限在真正使用的时候会向用户提示，用户的安全性和隐私得到保护，仅仅需要做一些适配工作，本节我们来解决两个问题。

如果设备运行的是 Android 6.0（API 级别 23）或更高版本，并且应用的 targetSdkVersion 是 23 或更高版本，则应用在运行时会向用户请求权限，用户可随时调用权限，因此应用在每次运行时均需检查自身是否具备所需的权限。

下面介绍使用知乎的一个例子（设备运行的是 Android 8.0 及更高版本），如图 7-1 所示；使用知乎的拍摄功能，依次会弹出相机权限、麦克风权限，用户运行程序获取权限，如果要取消这些权限，可以进入设置的应用权限管理，取消知乎的相机和麦克风权限。

图 7-1　知乎的权限实例

Android 将新的权限分为两类，一类是正常权限，比如联网、震动，这类权限跟之前一样，清单文件声明后直接授予，另一类是危险权限。手机常用软件的权限管理如图 7-2 所示。

图 7-2　手机常用软件的权限管理

从微信、360 手机助手、12306 订票软件的权限管理可以看到：读取联系人、存储、位置、电话、相机、短信、传感器、麦克风、相机、定位等涉及用户隐私的，需要在使用时通知用户进行授权。如图 7-3 所示的是 Android 需要申请的权限组。

图 7-3　Android 需要申请的权限组

Android 6.0 之前，用户安装 App，只需要把 App 需要的权限列出来告知用户，App 安装后都可以访问这些权限。Android 6.0 之后，敏感权限可以动态申请，用户可以拒绝；已授予的权限，可以在权限设置管理中关闭。

对于开发者来说，必须对权限管理做适配，否则 App 访问容易出现崩溃的现象。Android 的权限使用流程如图 7-4 所示，应用如果需要使用权限，首先应检查权限，如果已被授予权限，则可以直接使用权限；如果没有被授予权限，可以向用户请求权限，用户同意后方可使用权限。

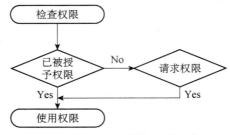

图 7-4　Android 的权限使用流程

拨打电话与发送短信（实践案例）

7.1.2　实践案例——设置电话权限

下面我们以一个电话权限的使用实例来讲解如何申请权限；如图 7-5 所示，单击打电话按钮（MAKE CALL），由于用户没有权限，弹出权限申请界面，用户同意后进入拨打电话界面。

图 7-5　拨打电话实例

拨打电话实例具体实现如下。

① 使用 checkSelfPermission() 检查权限。请求某个权限时首先应检查这个权限是否已经被用户授权，已经授权的权限重复申请可能会让用户产生厌烦；权限名称为 Manifest.permission.CALL_PHONE；函数返回值与 PackageManager.PERMISSION_GRANTED 作比较，一致则代表有权限，不一致代表没有权限，如图 7-6 所示。

```java
makeCall.setOnClickListener(new View.OnClickListener() {
    @Override
    public void onClick(View v) {  // 按键响应函数
        if (ContextCompat.checkSelfPermission( context: MainActivity.this,  // 判断应用是否有拨打电话权限
                Manifest.permission.CALL_PHONE)!=PackageManager.PERMISSION_GRANTED)
        {
            ActivityCompat.requestPermissions( activity: MainActivity.this,new String[]
                    {Manifest.permission.CALL_PHONE}, requestCode: 1);
        }
        else{  // 如果没有权限的话，申请权限
            call();
        }  // 如果有权限的话可以直接调用拨打函数的权限
    }
```

图 7-6　checkSelfPermission 检查权限

② 使用 requestPermissions()申请权限。调用后系统会显示一个请求用户授权的提示对话框，App 不能配置和修改这个对话框，如果需要提示用户这个权限相关的信息或说明，需要在调用 requestPermissions()之前处理，该方法有两个参数：

◆ int requestCode（参数需要圈记），会在回调 onRequestPermissionsResult()时返回，用来判断是哪个授权申请的回调。

◆ String[] permissions（参数需要圈记），权限数组，你需要申请的权限的数组。

由于该方法是异步的，所以无返回值，当用户处理完授权操作时，会回调 Activity 或者 Fragment 的 onRequestPermissionsResult()方法，如图 7-7 所示。

```java
public void onRequestPermissionsResult(int requestCode, String[] permissions, int[] grantResults) {
    switch (requestCode) {  // 用户申请的时候的协议要求，比如这次申请的权限标志码为1    Activity的系统重载函数
        case 1:
            if (grantResults.length > 0 && grantResults[0] == PackageManager.PERMISSION_GRANTED) {
                call();  // 调用拨打电话系统功能                              // 用户同意
            } else {
                Toast.makeText( context: this, text: "You denied the permission", Toast.LENGTH_SHORT).show();
            }
            // 用户不同意直接退出
            break;
        default:
    }
}
```

图 7-7　requestPermissions 申请权限代码

③ Activity 或者 Fragment 重写 onRequestPermissionsResult()方法处理权限结果回调，如图 7-8 所示；当用户处理完授权操作时，系统会自动回调该方法，该方法有 3 个参数。

◆ int requestCode：在调用 requestPermissions()时的第一个参数。

◆ String[] permissions：权限数组，在调用 requestPermissions()时的第二个参数。

◆ int[] grantResults：授权结果数组。

```java
public void onRequestPermissionsResult(int requestCode, String[] permissions, int[] grantResults) {
    switch (requestCode) {  // 用户申请的时候的协议要求，比如这次申请的权限标志码为1    Activity的系统重载函数
        case 1:
            if (grantResults.length > 0 && grantResults[0] == PackageManager.PERMISSION_GRANTED) {
                call();  // 调用拨打电话系统功能                              // 用户同意
            } else {
                Toast.makeText( context: this, text: "You denied the permission", Toast.LENGTH_SHORT).show();
            }
            // 用户不同意直接退出
            break;
        default:
    }
}
```

图 7-8　onRequestPermissionsResult 方法处理权限结果回调代码

④ 获取权限后使用系统 Intent.ACTION_CALL 值调用系统拨打电话功能，如图 7-9 所示。

```
private void call() {
    try {
        Intent intent = new Intent(Intent.ACTION_CALL);
        intent.setData(Uri.parse("tel:10086"));
        startActivity(intent);
    } catch (SecurityException e) {
        e.printStackTrace();
    }
}
```

图 7-9　onRequestPermissionsResult 处理权限结果

7.1.3　单元小测

判断题：

1. Android 6.0 后，App 将在安装的时候授予权限。（　　）
   A. 是　　　　　　　　　　　　B. 否
2. Android 6.0 后，App 将在运行的时候向用户申请权限。（　　）
   A. 是　　　　　　　　　　　　B. 否
3. 对于开发者来说，不需要权限管理做适配，App 访问权限不会出现崩溃的现象。（　　）
   A 是　　　　　　　　　　　　B. 否
4. 对于已授予的权限，不可以通过权限设置管理去关闭。（　　）
   A. 是　　　　　　　　　　　　B. 否

选择题：

1. Android 常用的权限有哪些？（　　）
   A. 手机存储　　　B. 通讯录　　　C. 短消息　　　D. 通话记录
2. 运行权限申请中 requestCode 的作用是（　　）。
   A. 用户申请权限的校验码
   B. 用户申请权限的标志码
   C. 无任何作用
   D. 用户申请后系统返回给用户的标志码

7.2　URL 和 URI 概述

7.2.1　URL 和 URI

URL 和 URI 概述
（慕课）

本节我们介绍 URI 和 URL，URI（Uniform Resource Identifier）是统一资源标识符，用于标识一个抽象或者物理资源；URL（Uniform Resource Locator）是资源定位符，用于标识网络资源的位置；URI 是以一种抽象的高层次概念定义统一资源标志，而 URL 则是具体的资源标志的方式。URL 是一种特殊的 URI。

URI 一般由下面三部分组成。
- scheme：组件，资源的名称标志。
- scheme-specific-part：存放资源的主机名，表示唯一标志。
- Fragment：子资源，由路径表示。

URL 的格式一般由下列三部分组成。
- scheme：协议（或称为服务方式）。
- scheme-specific-part：存有该资源的主机 IP 地址（有时也包括端口号）。
- Fragment：主机资源的具体地址。

URL 的具体例子如下所示。
- HTTP 协议：http://www.csg.com/index.html#28，代表 HTTP 网络访问地址。
- FTP 协议：ftp://212.23.20.128/manager/user.zip，代表 FTP 文件访问地址。
- Content 文件访问协议：content://com.android.contacts/raw_contacts，代表 Content 文件访问地址。

URI 与 URL 的区别如下：
- URI 表示资源的识别符号，从某种意义来看，它只是一个符号而已。
- URL 称为统一资源定位，强调的是定位，在 URL 中一定包含定位信息，比如服务器地址之类的，相比较而言 URI 中未必要包含定位信息。
- URL 符合 URI 的组成规则，可以看成是一种特殊的 URI。

Android 上可用的每种资源如图像、视频片段等都可以用 URI 来表示。Android 的 URI 由以下三部分组成："content://"、数据的路径、标志 ID。

Android URI 实例如下所示。
- 所有联系人的 URI：content://contacts/people。
- 某个联系人的 URI：content://contacts/people/5。
- 所有图片 URI：content://media/external。
- 某个图片的 URI：content://media/external/images/media/4。

Android 上多媒体 URI 属性如表 7-1 所示。

表 7-1 多媒体 URI 属性

序号	多媒体名称	多媒体 URI
1	SD 卡上的音频文件	MediaStore.Audio.Media.EXTERNAL_CONTENT_URI
2	手机内部存储器音频文件	MediaStore.Audio.Media.INTERNAL_CONTENT_URI
3	SD 卡上的图片文件	MediaStore.Images.Media.EXTERNAL_CONTENT_URI
4	手机内部存储器上的图片	MediaStore.Images.Media.INTERNAL_CONTENT_URI
5	SD 卡上的视频	MediaStore.Video.Media.EXTERNAL_CONTENT_URI
6	手机内部存储器上的视频	MediaStore.Video.Media.INTERNAL_CONTENT_URI

7.2.2 单元小测

判断题：

1. URI 的全称为 Uniform Resource Identifier。（   ）

A. 是 B. 否

2. URI 一定包含定位信息。(　　)

A. 是 B. 否

3. URL 一定包含定位信息，可以看成是一种特殊的 URI。(　　)

A. 是 B. 否

选择题：

1 下面哪些协议是符合 URI 标识的组成规则？(　　)

A. HTTP　　　　　B. SMS　　　　　C. FTP　　　　　D. Content

2 下面哪些是符合 URI 标识的组成规则？(　　)

A. http://www.csg.com/index.html#2

B. ftp://212.23.20.128/manager/user.zip

C. content://com.android.contacts/raw_contacts

D. sms://com.example.app.provider/table1

7.3 ContentProvider

7.3.1 知识点讲解——ContentProvider

ContentProvider 是 Android 的四大组件之一，ContentProvider 一般为存储和获取数据提供统一的接口，可以在不同的应用程序之间共享数据。ContentProvider 的架构如图 7-10 所示。

ContentProvider 提供了对底层数据存储方式的抽象。底层使用了 SQLite 数据库，在使用了 ContentProvider 封装后，即使你把数据库换成 MongoDB，也不会对上层代码产生影响。

Android 框架中的一些类需要 ContentProvider 类型数据。如果你想让你的数据可以使用在如 SyncAdapter、Loader、CursorAdapter 等类上，那么你就需要为你的数据做一层 ContentProvider 封装。

ContentProvider 为应用间的数据交互提供了一个安全的环境。它准许你把自己的应用数据根据需求开放给其他应用进行增、删、改、查操作，而不用担心直接开放数据库权限而带来的安全问题。

ContentProvider 是对数据层的封装，我们可以通过 ContentResolver 来对不同的 ContentProvider 进行增、删、改、查的操作，如图 7-11 所示。

图 7-10 ContentProvider 的架构

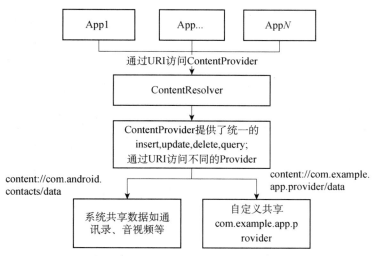

图 7-11 ContentResolver 接口架构

Android 为常见的一些数据提供了默认的 ContentProvider（包括音频、视频、图片和通讯录等），所以我们可以在其他应用中通过 URI 获取这些数据，比如通过 content://com.android.contacts/data 来获取通讯录数据。

ContentProvider 中的 URI 有固定格式，如图 7-12 所示。

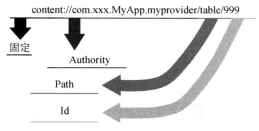

图 7-12 URI 的固定格式

- ◆ Authority：授权信息，用以区别不同的 ContentProvider。
- ◆ Path：表名，用以区分 ContentProvider 中不同的数据表。
- ◆ Id：ID 号，用以区别表中的不同数据。

如果是自定义的数据，其他应用程序可以通过自定义共享的 URI 标志 com.example.app.provider 来访问数据；为什么我们不直接访问 Provider，而是又在上面加了一层 ContentResolver 来进行操作，这样岂不是更复杂了吗？如图 7-13 所示，大家要知道一个手机中可不是只有一个 Provider 内容，它可能安装了很多含有 Provider 的应用，比如联系人应用、日历应用、字典应用等。所以 Android 提供了 ContentResolver 来统一管理不同 ContentProvider 间的操作。

ContentProvider 要实现的方法如下：onCreate 方法用于初始化 Provider；query 方法用于提供数据查询能力；insert 方法用于提供增加数据的能力；update 方法用于提供更新数据的能力；delete 方法用于提供删除数据的能力；getType 方法用于返回 ContentProvider 中的数据类型。

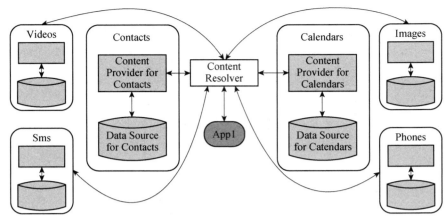

图 7-13　ContentResolver 接口操作

下面以 query 为例子讲解调用的接口。query 的接口 query(uri,projection,selection,selectionArgs,sortOrder)参数如表 7-2 所示。

表 7-2　query 的接口参数

序号	参数	说明	描述
1	uri	from table_name	通过 URI 指定查询应用程序的表
2	projection	select column1,column2	指定应用程序表的列名
3	selection	where column=value	指定列名中的元素字段的约束条件
4	selectionArgs	new String[]{name,…}	为 where 中的占位符提供具体的值
5	sortOrder	order by column1,column2	指定查询结果的排序方式

读取联系人
（实践案例）

7.3.2　实践案例——读取联系人

下面以一个通讯录的实例讲解参数的具体使用方法，如图 7-14 所示。

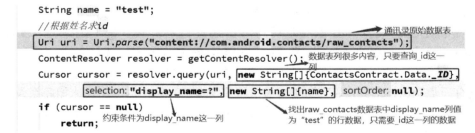

图 7-14　访问通讯录的参数

- uri:通讯录的原始数据表。
- projection:通讯录数据列表有很多内容，只需要查询_ID 这一列。
- selection:约束条件为 display_name 这一列。
- selectionArgs:从通讯录数据列表中查询 display_name 列值为"test"的行数据,只需要_id

这一列的数据。
- sortOrder:指定查询结果的排序方式为默认。

Android Sdk 中提供的 ContentProvider 接口大概有 10 类，如图 7-15 所示。以联系人为例，数据存储位置为/data/data/com.android.providers.contacts/database，其中 contact2.db 中存储了联系人的基本信息。

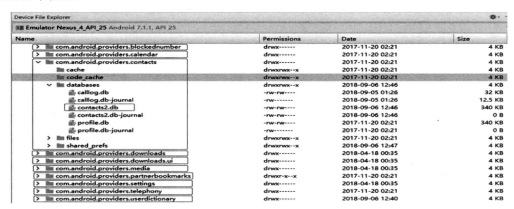

图 7-15　ContentProvider 接口

"contacts2.db"主要有 4 个表：contacts、data、raw_contacts、mimetypes。

① contacts 表简单存储了联系人的一些我信息，如图 7-16 所示。

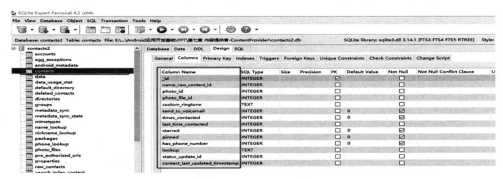

图 7-16　contacts 表

外键字段 name_raw_contact_id 对应着表 raw_contacts 表中的字段_id，如图 7-17 所示。

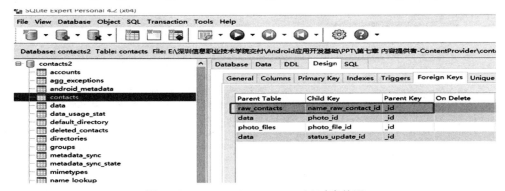

图 7-17　name_raw_contact_id 对应关系

② data 表简单存储了联系人的详细数据，如图 7-18 所示。

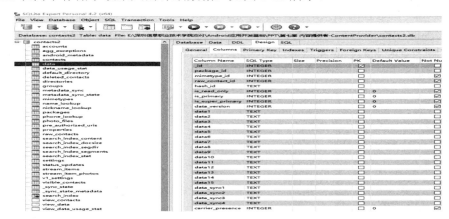

图 7-18　data 表

外键字段 raw_contact_id 对应着表 raw_contacts 表中的字段 _id，如图 7-19 所示。

图 7-19　raw_contact_id 对应关系

③ raw_contacts 表是联系人存储数据的核心表，如图 7-20 所示。

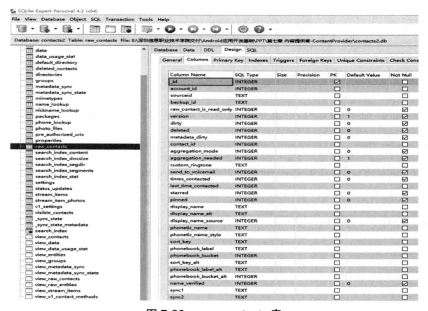

图 7-20　raw_contacts 表

④ mimetypes 表：表中存储着与联系人相关的数据分类，主要包括 email，phone，im 等，如图 7-21 所示。

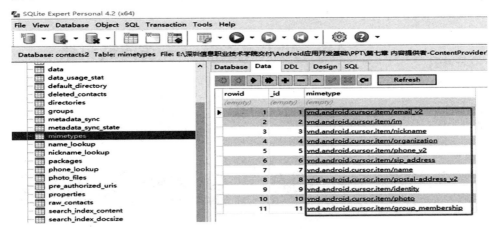

图 7-21　mimetypes 表

data 表中的数据分类主要包括联系人的常用数据，如图 7-22 所示。

图 7-22　data 表数据

7.3.3　单元小测

判断题：

1. ContentProvider 所提供的 URI 可以随便定义。（　　）
　A. 是　　　　　　　　　　　　　B. 否
2. ContentResolver 可以通过 ContentProvider 所提供的 URI 进行数据操作。（　　）
　A. 是　　　　　　　　　　　　　B. 否
3. ContentProvider 操作数据的时候，必须在配置信息文件中注册。（　　）
　A. 是　　　　　　　　　　　　　B. 否

选择题：

1. 系统可以通过 ContentProvider 进行下列哪些数据的共享？（　　）
　A. 音频数据　　　B. 通话记录数据　　　C. 系统内存数据　　　D. 短消息数据
2. 在实现 ContentProvider 时需要实现的方法包括哪些？（　　）
　A. Query 方法　　B. Update 方法　　　C. Delete 方法　　　D. Insert 方法

7.4 访问通讯录

7.4.1 知识点讲解——访问通讯录

访问通讯录（慕课）

在 Android 中，可以使用 ContentResolver 对通讯录中的数据进行添加、删除、修改和查询操作。如图 7-23 所示的是读取联系人信息的实例，首先向用户获取读取通讯录的权限，用户同意后，读取用户的通讯录信息并显示。

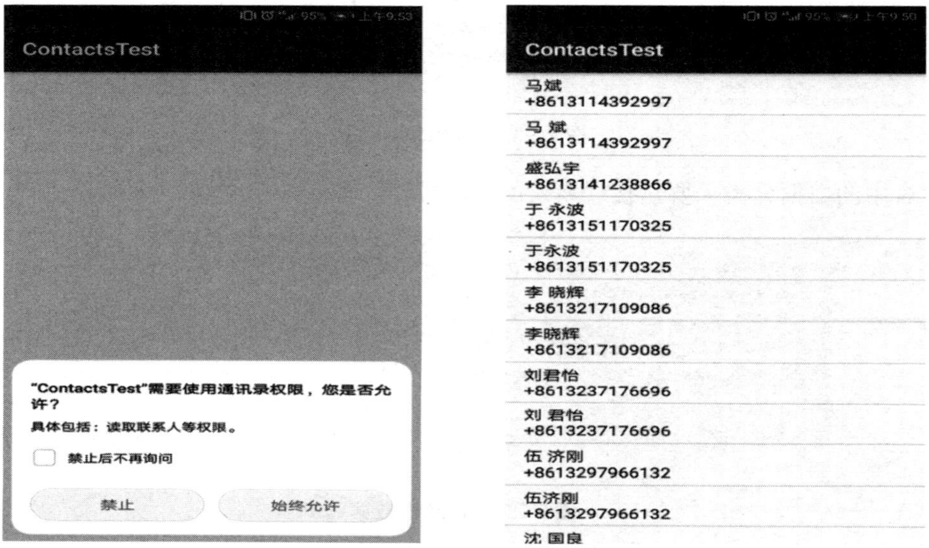

图 7-23 读取联系人信息的实例

在查询联系人时应该先查询 raw_contacts 中的数据，对应的 URI 分别为 ContactsContract.CommonDataKinds.Phone.CONTENT_URI；URI 实际值为 content://com.android. contacts/data/phones（除 ppt 对应内容）；URI 对应 data、contacts、raw_contacts 这三张表。

使用 getContentResolver 获得 ContentResolver 的对象：ContentResolver mCr= context. getContentResolver()。

通过 ContentResolver 查询 URI，得到数据库游标 Cursor c：Cursor c = mCR.query **(ContactsContract.CommonDataKinds.Phone.CONTENT_URI**, null, null, null, null)。

使用数据库游标 Cursor 获取联系人的姓名和手机号，代码如图 7-24 所示。

```
// 获取联系人姓名
String displayName = cursor.getString(cursor.getColumnIndex(ContactsContract.CommonDataKinds.Phone.DISPLAY_NAME));
// 获取联系人手机号
String number = cursor.getString(cursor.getColumnIndex(ContactsContract.CommonDataKinds.Phone.NUMBER));
```

图 7-24 获取联系人的姓名和手机号代码

7.4.2 实践案例——访问通讯录

下面我们看一下具体的实现过程,首先进行通讯录权限的判断和申请,如图 7-25 所示。

检查应用是否有读取通讯录的权限,如果没有权限的话需要向用户申请权限,有权限的话直接调用读取通讯录函数。

读取通讯录
(实践案例)

```
if (ContextCompat.checkSelfPermission( context: this,
        Manifest.permission.READ_CONTACTS) !=
        PackageManager.PERMISSION_GRANTED)
{
    ActivityCompat.requestPermissions( activity: this,
            new String[]{Manifest.permission.READ_CONTACTS},
            requestCode: 1);
} else {
    readContacts();
}
```

检查应用是否有读取通讯录权限

没有权限的话需要申请权限

如果有权限的话直接调用读取通讯录函数

图 7-25 通讯录权限的判断和申请

权限申请后的处理,如图 7-26 所示,Activity 或者 Fragment 重写 onRequestPermissionsResult() 方法处理权限结果回调;当用户处理完授权操作时,系统会自动回调该方法,该方法有三个参数。

◆ int requestCode:在调用 requestPermissions()时的第一个参数。
◆ String[] permissions:权限数组,在调用 requestPermissions()时的第二个参数。
◆ int[] grantResults:授权结果数组。

```
public void onRequestPermissionsResult
        (int requestCode, String[] permissions, int[] grantResults) {
    switch (requestCode) {
        case 1:
            if (grantResults.length > 0 && grantResults[0] ==
                    PackageManager.PERMISSION_GRANTED) {
                readContacts();
            } else {
                Toast.makeText( context: this,
                        text: "You denied the permission", Toast.LENGTH_SHORT).show
            }
            break;
```

图 7-26 通讯录权限申请回调处理

读取联系人信息数据中的姓名和号码,如图 7-27 所示。首先获取系统的 ContentResolver,使用 ContentResolver 的 query 接口查询系统的电话通讯录数据,如果查询到数据,逐条查询通讯录数据库中的数据,查询每一条记录的姓名和号码,将名字和号码信息保存到列表数组中。

图 7-27 读取联系人信息

代码编写完成后看一下运行的效果,如图 7-28 所示,首先提示需要使用通讯录权限,如果用户允许后,读取通讯录并显示。

图 7-28 读取联系人信息运行效果

如何插入联系人呢?首先看一下数据库 data 表,我们要插入一条数据到通讯录数据库,需要使用 ContentResolver 的 insert 接口,数据库 data 表如图 7-29 所示。

图 7-29 插入联系人前的 data 表

需要使用 uri 查询需要添加的号码是否已存入通讯录中,如果不存在则添加,存在则提示用户。判断需要添加的号码已存入通讯录的代码如图 7-30 所示。

```
//先查询要添加的号码是否已存在通讯录中，不存在则添加，存在则提示用户
Uri uri = Uri.parse("content://com.android.contacts/data/phones/" ·
        "filter/" + number);
ContentResolver resolver = getContentResolver();
//从raw_contact 表中返回display_name
Cursor cursor = resolver.query(uri, new String[]{ContactsContract.
                Data.DISPLAY_NAME},
        selection: null, selectionArgs: null, sortOrder: null);
if (cursor == null)
    return;
```

图 7-30　判断需要添加的号码已存入通讯录的代码

向 data 表中插入数据，如图 7-31 所示，为 uri 对象赋值，使用 value.put 分别将数据写入 "raw_contact_id" "MIMETYPE" "data2" "data1" 4 列中，使用 ContentResolver 的 insert 接口插入到通讯录数据库，插入数据的代码及插入后的数据库如图 7-31 所示。

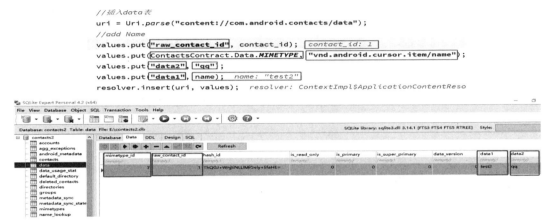

图 7-31　插入数据的代码及插入后的数据库

使用同样的方法向 data 表中插入数据；使用 value.put 分别将数据写入 "raw_contact_id" "MIMETYPE" "data2" "data1" 4 列中，使用 ContentResolver 的 insert 接口插入到通讯录数据库，插入后的数据库如图 7-32 所示。

删除和更改通讯录
（实践案例）

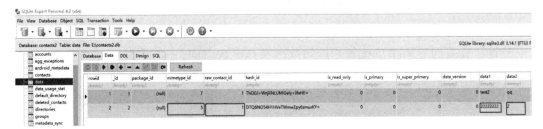

图 7-32　插入后的数据库

如何删除联系人呢？首先查看一下数据库 raw_contacts 表，如图 7-33 所示，数据库表中有_id 和 display_name 这两列，我们需要根据 display_name 这一列的约束条件为 test 得到_id

这一列的值。

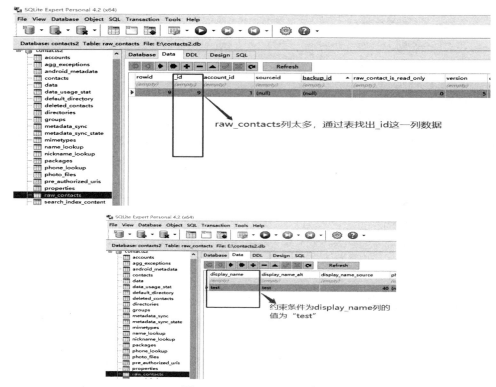

图 7-33 raw_contacts 表

首先将 uri 设置为通讯录的原始数据表 raw_contacts，使用 ContentResolver 的 query 接口找出 raw_contacts 数据表中 display_name 这一列值为"test"的行数据，只取"_id"这一列的数据，查找的代码如图 7-34 所示。

图 7-34 查找的代码

下面我们看一下如何删除数据，如图 7-35 所示，在 data 表中，raw_contact_id 等于 9 的数据有很多行，需要同时删除 data 表和原始数据表 raw_contacts 中的数据。

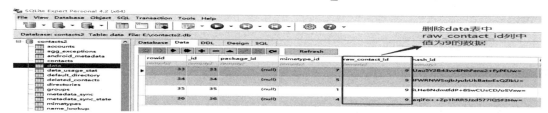

图 7-35 删除前的 data 表

查询成功后读取 raw_contacts 表中满足条件的行数据的_id 值；根据 display_name 的值删除 raw_contacts 表中的相应数据；根据_id 值删除 data 表中 raw_contact_id 值等于 9 的相应数据；实现代码如图 7-36 所示。

图 7-36　删除数据代码

如何更改联系人呢？首先查看一下数据库 raw_contacts 表，如图 7-37 所示，数据库表中有_id 和 display_name 这两列；我们需要根据 display_name 这一列的约束条件为 test 得到_id 这一列的值。

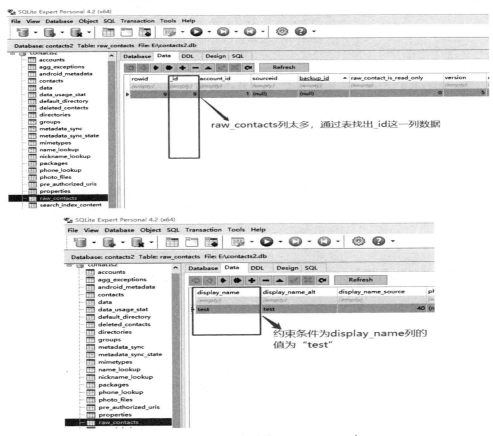

图 7-37　更改联系人前的 raw_contacts 表

首先将 uri 设置为通讯录的原始数据表 raw_contacts；使用 ContentResolver 的 query 接口找出 raw_contacts 数据表中 display_name 这一列值为"test"的行数据，只取"_id"这一列的数据，查找的代码如图 7-38 所示。

```
String name = "test";
//根据姓名求id
Uri uri = Uri.parse("content://com.android.contacts/raw_contacts");
ContentResolver resolver = getContentResolver();
Cursor cursor = resolver.query(uri, new String[]{ContactsContract.Data._ID},
        selection: "display_name=?", new String[]{name}, sortOrder: null);
if (cursor == null)
    return;
```

通讯录原始数据表

数据表列很多内容，只要查询_id这一列

约束条件为display_name这一列

找出raw_contacts数据表中display_name列值为"test"的行数据，只需要_id这一列的数据

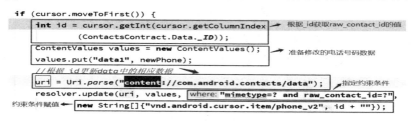

图 7-38 查询找的代码

下面我们来实现更改数据，在 data 表中，raw_contact_id 等于 9 的数据有很多行。需要更新的是 mimetype 为 5 的也就是电话号码的数据，如图 7-39 所示的是更新通讯录的代码。

```
if (cursor.moveToFirst()) {
    int id = cursor.getInt(cursor.getColumnIndex
            (ContactsContract.Data._ID));
    ContentValues values = new ContentValues();
    values.put("data1", newPhone);
    //根据_id更新data中的相应数据
    uri = Uri.parse("content://com.android.contacts/data");
    resolver.update(uri, values, where: "mimetype=? and raw_contact_id=?",
            new String[]{"vnd.android.cursor.item/phone_v2", id + ""});
```

根据_id获取raw_contact_id的值

准备修改的电话号码数据

指定约束条件

约束条件赋值

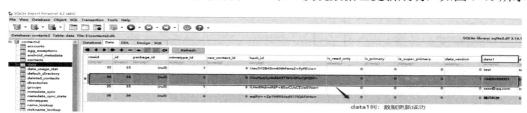

图 7-39 更新通讯录的代码

查询成功后读取 raw_contacts 表中满足条件的行数据的_id 的值为 9，使用 values.put 将要修改的电话号码放入列 data1 中，根据_id 的值更新 data 表中 mimetype 等于 5 并且 raw_contact_id 值等于 9 的相应数据；最后我们查看数据库 data 表，发现数据已更新成功，如图 7-40 所示。

图 7-40 data 表更新成功

7.4.3 单元小测

判断题：

1. ContentResolver 可以操作数据库中的数据。（　　）

A. 是　　　　B. 否

2. ContentResolver 可以操作 ContentProvider 暴露的数据。（　　）

A. 是　　　　B. 否

3. ContentResolver 操作其他应用数据必须知道包名。（　　）

A. 是　　　　B. 否

选择题：

1. 获取系统的 ContentResolver 一般使用什么方法？（　　）

A. getContentProvider()　　　　B. getSystemContentProvider()

C. getSystemContentResolver()　　　　D. getContentResolver()

2. 在 ContentResolver 中查询系统的数据一般使用什么方法？（　　）
A. query　　　　　　B. update　　　　　　C. delete　　　　　　D. insert
3. 在 ContentResolver 中删除系统的数据一般使用什么方法？（　　）
A. query　　　　　　B. update　　　　　　C. delete　　　　　　D. insert
4 在 ContentResolver 中更新系统的数据一般使用什么方法？（　　）
A. query　　　　　　B. update　　　　　　C. delete　　　　　　D. insert
5 在 ContentResolver 中向系统新增数据一般使用什么方法？（　　）
A. query　　　　　　B. update　　　　　　C. delete　　　　　　D. insert

本章课后练习和程序源代码

第 7 章源代码及课后习题

8 多媒体

Android 的拍照服务、Android 的相册服务、Android 的音视频服务。

1. 熟练应用 Android 的拍照服务。
2. 熟练访问 Android 的相册服务。
3. 熟练使用 Android 的音视频服务。

8.1 拍照服务

8.1.1 知识点讲解——拍照服务

本节我们介绍 Android 的拍照服务。现在的 Android 智能手机都会提供拍照的功能，大部分手机的摄像头的像素都在 1000 万以上，有的甚至会更高。它们大多都会支持光学变焦、曝光及快门等。大部分的应用中，图片分享已成为主要的信息载体，比如微信的朋友圈、QQ 的图片分享、微博的图片分享等，如图 8-1 所示。

相机服务
（慕课）

相机也需要使用 Android 的系统服务，如图 8-2 所示。Android 系统为了降低系统服务的编程难度，研发了大量 Manager 组件。这些独立的管理组件可以简化访问系统服务的难度。我们主要通过 Photo Manager 去访问系统的拍照和相册服务。

图 8-1 相机应用场景

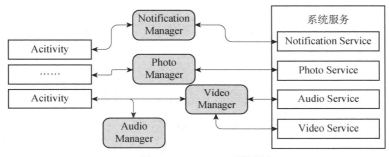

图 8-2 Android 系统服务

8.1.2 实践案例——拍照服务

下面我们看一个相册的使用实例：完成一个包含"拍摄"和"从相册中选择"两个按钮的应用；单击"拍摄"按钮，使用相机拍摄照片；获得用户存储权限后显示刚拍摄的照片；单击"从相册中选择"按钮，从手机相册中选择一张图片并显示，如图 8-3 所示。

拍照服务
（实践案例）

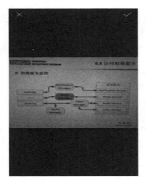

图 8-3 Android 相机使用实例

下面看一下拍照服务如何实现。首先确定文件路径，如图 8-4 所示，创建 File 对象，用于存储拍摄照片，使用 getExternalCacheDir 获取 SD 卡路径，将拍摄照片命名为 output.jpg。

低于 Android 7.0 的版本，可以直接将 File 对象转换为 URI 对象；高于 Android 7.0 的版本，不能直接使用 URI，需要使用 FileProvider 将 URI 封装后共享给外部对象。

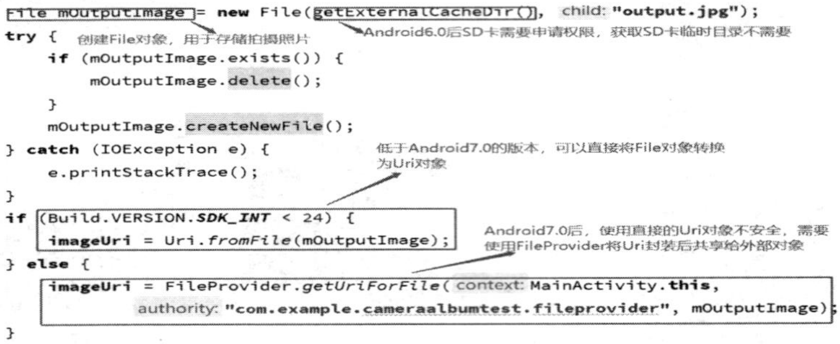

图 8-4　确定文件路径代码

启动相机代码如图 8-5 所示，新建一个相机系统服务的意图；为系统服务传递文件路径的 URI 地址；使用 startActivityForResult 开启拍照服务，并要求相机拍照服务完成后将返回值给视图。

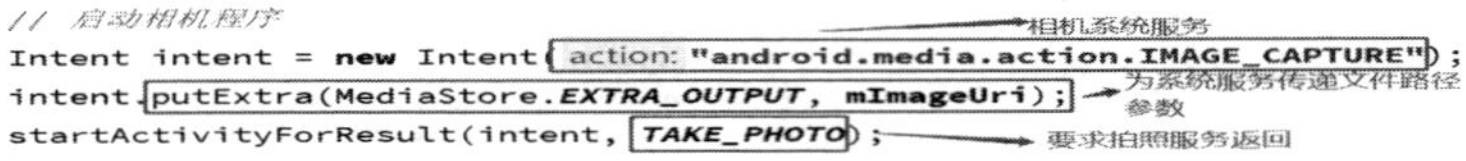

图 8-5　启动相机代码

在 Activity 中重写 onActivityResult()方法处理相册系统服务结果回调，如图 8-6 所示；当拍照服务完成后，系统会自动回调该方法。拍照服务的图片放入用户传入的 URI 地址中；将拍摄的照片显示出来。

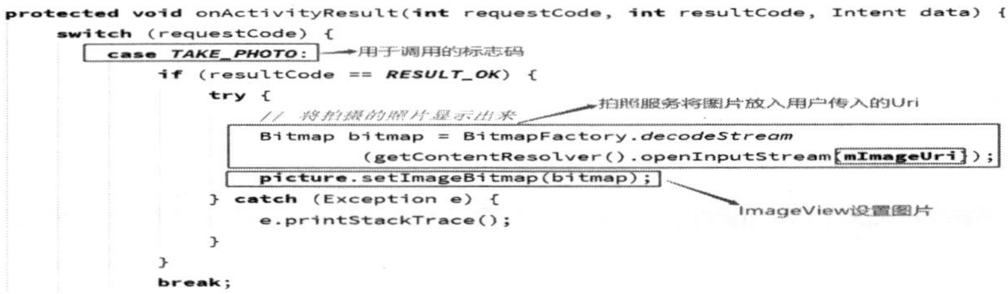

图 8-6　onActivityResult 处理回调代码

FileProvider 注册代码如图 8-7 所示；在配置信息文件中新建一个 provider 标签；android:name 值是固定的；android:authorities 的值与 FileProvider 中 getUriForFile 的值保持一致；URI 的共享路径设置为@xml/file_paths。

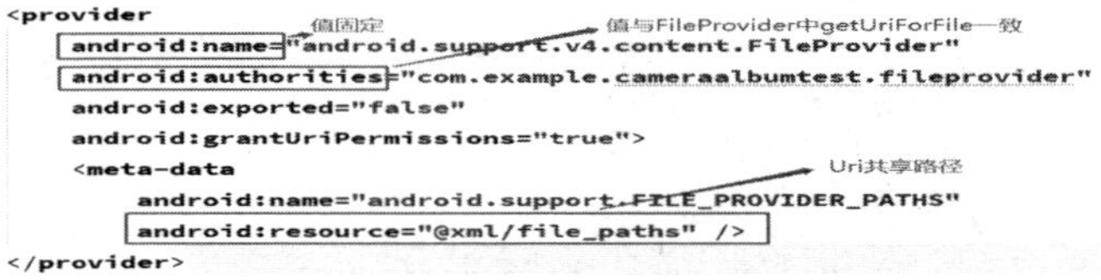

图 8-7　FileProvider 注册代码

URI 共享路径在 res/xml 目录中新建 file_paths 文件，如图 8-8 所示。设置外部路径 external-path name="my_images"。

```
<paths xmlns:android="http://schemas.android.com/apk/res/android">
    <external-path name="my_images" path="" />
</paths>
```

图 8-8　URI 共享路径代码

在配置文件中设置用户的 SD 卡权限，如图 8-9 所示，这样我们就完成了所有的拍照功能。

```
<uses-permission android:name="android.permission.WRITE_EXTERNAL_STORAGE" />
```

图 8-9　配置 SD 卡权限代码

下面看一下访问相册服务该如何实现。首先需要申请 SD 卡读写权限，使用 checkSelfPermission() 检查权限；请求某个权限时检查这个权限是否已经被用户授权，已经授权的权限重复申请可能会让用户产生厌烦；权限名称为 Manifest.permission.WRITE_ EXTERNAL_STORAGE；函数返回值与 PackageManager.PERMISSION_GRANTED 作比较，一致则代表有权限，不一致代表没有权限；使用 requestPermissions() 申请权限；由于该方法是异步的，所以无返回值，当用户处理完授权操作时，会回调 Activity 的 onRequestPermissionsResult() 方法，如图 8-10 所示。

相册服务
（实践案例）

```
if (ContextCompat.checkSelfPermission( context: MainActivity.this,    检查是否有SD卡写权限
        Manifest.permission.WRITE_EXTERNAL_STORAGE) != PackageManager.PERMISSION_GRANTED)
{
        ActivityCompat.requestPermissions( activity: MainActivity.this,    如果没有权限申请SD卡写权限
                new String[]{ Manifest.permission. WRITE_EXTERNAL_STORAGE }, requestCode: 1);
} else {
    openAlbum();
}
```

图 8-10　申请 SD 卡读写权限代码

Activity 重写 onRequestPermissionsResult() 方法处理权限结果回调；当用户处理完授权操作时，系统会自动回调该方法，如果用户同意则打开相册，如图 8-11 所示。

```
public void onRequestPermissionsResult(int requestCode, String[] permissions, int[] grantResults) {
    switch (requestCode) {
        case 1:
            if (grantResults.length > 0 && grantResults[0] ==
                    PackageManager.PERMISSION_GRANTED) {
                openAlbum();    如果用户同意打开相册
            } else {
                Toast.makeText( context: this, text: "You denied the permission",
                        Toast.LENGTH_SHORT).show();
            }
            break;
        default:
    }
}
```

图 8-11　onRequestPermissionsResult 方法代码

完成打开系统相册功能，如图 8-12 所示。新建一个访问相册系统服务的意图，为系统服务设置访问文件类型；使用 startActivityForResult 开启拍照服务，并要求拍照服务完成后将返回值传给视图。

```
private void openAlbum() {                                        相册系统服务
    Intent intent = new Intent( action: "android.intent.action.GET_CONTENT");
    intent.setType("image/*");            设置访问文件类型
    startActivityForResult(intent, CHOOSE_PHOTO);    打开拍照服务（带返回值的）
}                                                    打开相册
```

图 8-12　打开系统拍照功能代码

在 Activity 中重写 onActivityResult() 方法处理拍照服务结果回调；根据手机系统版本号处理返回的数据；Android 4.4 及以上的系统使用 handleImageOnKitKat(data)处理数据；Android 4.4 以下的系统使用 handleImageBeforeKitKat(data)处理数据，如图 8-13 所示。

```java
protected void onActivityResult(int requestCode, int resultCode, Intent data) {
    switch (requestCode) {
        case CHOOSE_PHOTO:
            if (resultCode == RESULT_OK) {
                // 判断手机系统版本号
                if (Build.VERSION.SDK_INT >= 19) {
                    //4.4 及以上系统使用这个方法处理图片
                    handleImageOnKitKat(data);
                } else {
                    //4.4 以下系统使用这个方法处理图片
                    handleImageBeforeKitKat(data);
                }
            }
            break;
        default:
            break;
    }
}
```

图 8-13　onActivityResult 方法代码

Android 4.4 及以上的系统使用 handleImageOnKitKat(data)处理数据的流程比较复杂，如图 8-14 所示；根据返回值获取文件的 URI 路径；再根据 URI 来判断文件是否为文档；如果 URI 显示为本机的媒体文件，直接获取文件的路径；如果是下载文件，获取文件的下载路径；如果是 Content 类型文件，直接使用 getImagePath 获取文件的路径；如果是 File 类型文件，使用 URI 直接获取图片文件路径；最后根据图片文件路径显示图片。

```java
private void handleImageOnKitKat(Intent data) {
    String imagePath = null; Uri uri = data.getData();  //获取文件的Uri路径
    if (DocumentsContract.isDocumentUri(context: this, uri)) {  //根据Uri判断是否路径文件是否为文档
        String docId = DocumentsContract.getDocumentId(uri);
        if("com.android.providers.media.documents".equals(uri.getAuthority())) {  //是否为本机媒体文件
            String id = docId.split(regex: ":")[1];  // 解析出数字格式的id
            String selection = MediaStore.Images.Media._ID + "=" + id;      //获取文件过滤路径
            imagePath = getImagePath(MediaStore.Images.Media.EXTERNAL_CONTENT_URI, selection);
        } else if ("com.android.providers.downloads.documents".
                equals(uri.getAuthority())) {                               //是否为下载文件
            Uri contentUri = ContentUris.withAppendedId(Uri.parse           //获取文件
                    ("content://downloads/public_downloads"), Long.valueOf(docId));  //的下载路径
            imagePath = getImagePath(contentUri, selection: null);
        }
    } else if ("content".equalsIgnoreCase(uri.getScheme())) {               //如果是Content类型，普通方式处理
        imagePath = getImagePath(uri, selection: null);
    } else if ("file".equalsIgnoreCase(uri.getScheme())) {                  //如果是File类型，使用Uri直接获取
        imagePath = uri.getPath();                                          //图片文件路径
    }
    displayImage(imagePath); // 根据图片路径显示图片                          //根据图片文件路径显示图片
}
```

图 8-14　handleImageOnKitKat 方法代码

Android 4.4 及以下的系统对图片处理的方式比较简单，如图 8-15 所示。根据返回值获取图片文件的 URI 路径；再根据 URI 获取图片文件的路径；最后根据图片文件的路径显示图片。

```java
private void handleImageBeforeKitKat(Intent data) {
    Uri uri = data.getData();
    String imagePath = getImagePath(uri, null);
    displayImage(imagePath);
}
```

图 8-15　handleImageBeforeKitKat 方法代码

实现根据 URI 地址获取图片文件的地址采用 getImagePath 方法，其代码如图 8-16 所示；首先通过 URI 查询图片信息；再轮询数据库指针，通过 cursor 的 getColumnIndex 方法查找 MediaStore.Images.Media.DATA 列的索引号，最后根据索引号获取图片的存放地址。

```java
private String getImagePath(Uri uri, String selection) {
    String path = null;
    // 通过 Uri 和 selection 来获取真实的图片文件路径
    Cursor cursor = getContentResolver().query(uri, null, selection, null, null);
    if (cursor != null) {
        if (cursor.moveToFirst()) {
            path = cursor.getString(cursor.getColumnIndex(MediaStore.Images.Media.DATA));
        }
        cursor.close();
    }
    return path;
}
```

图 8-16　getImagePath 方法代码

根据图片文件的地址显示图片方法的 displayImage 代码，如图 8-17 所示；如果图片文件路径不为空，使用 Bitmap 将图片解码为位图文件；否则提示不能加载此图片。

```java
private void displayImage(String imagePath) {
    if (imagePath != null) {
        Bitmap bitmap = BitmapFactory.decodeFile(imagePath);
        mImgPhoto.setImageBitmap(bitmap);
    } else {
        Toast.makeText(this, "failed to get image", Toast.LENGTH_SHORT).show();
    }
}
```

图 8-17　displayImage 方法代码

8.1.3　单元小测

判断题：

1. PhotoManager 是 Android 中的一个系统服务组件。（　　）
A. 是　　　　　　　　　　　　　　　　B. 否
2. 拍照服务不需要在配置信息文件中申请权限。（　　）
A. 是　　　　　　　　　　　　　　　　B. 否

选择题：

1. 相机系统服务的值是下面的哪一个值？（　　）

A. android.media.action.PHOTO

B. android.media.action.IMAGE_CAPTURE

C. android.media.action.GET_CONTENT

D. android.media.action.ALBUM

2. 相册系统服务的值是下面的哪一个值？（　　）

A. android.media.action.PHOTO B. android.media.action.IMAGE_CAPTURE
C. android.media.action.GET_CONTENT D. android.media.action.ALBUM

3. 在配置信息文件中，SD 卡的读权限是下面的哪一个值？（   ）

A. android.permission.WRITE_STORAGE
B. android.permission.WRITE_EXTERNAL_STORAGE
C. android.permission.READ_EXTERNA_STORAGE
D. android.permission.READ_EXTERNAL_STORAGE

4. 在配置信息文件中，SD 卡的写权限是下面的哪一个值？（   ）

A. android.permission.WRITE_STORAGE
B. android.permission.WRITE_EXTERNAL_STORAGE
C. android.permission.READ_EXTERNA_STORAGE
D. android.permission.READ_EXTERNAL_STORAGE

5. 阅读下面代码

```
{
    Intent intent = new Intent(""android.media.action.IMAGE_CAPTURE"");
    intent.putExtra(MediaStore.EXTRA_OUTPUT, mImageUri);
    startActivityForResult(intent, TAKE_PHOTO);
};
```

代码中 TAKE_PHOTO 作用是什么？（   ）

A. 用户申请访问相册的标志码
B. 用户申请相机的标志码
C. 用户申请访问相册后系统返回给用户的标志码
D. 用户申请相机后系统返回给用户的标志码

6. 阅读下面代码

```
{
    Intent intent = new Intent(""android.intent.action.GET_CONTENT"");
    intent.setType(""image/*"");
    startActivityForResult(intent, CHOOSE_PHOTO);
};
```

代码中 CHOOSE_PHOTO 作用是什么？（   ）

A. 用户申请访问相册的标志码
B. 用户申请相机的标志码
C. 用户申请访问相册后系统返回给用户的标志码
D. 用户申请相机后系统返回给用户的标志码

音视频服务（慕课）

 8.2 音视频服务

8.2.1 知识点讲解——音视频服务

本节我们介绍 Android 的音视频服务。现在的 Android 智能手机都会提供音视频的功能。大部分的应用中，视频分享已成为主要的信息载体。2020 年，视频将占到数据中心和终端用户传输总流量的 85%，比如抖音、快手等都是用户量非常大的视频应用，如图 8-18 所示。

图 8-18　视频应用案例

音视频也需要使用 Android 的系统服务，如图 8-19 所示。我们主要通过 Video Manager 和 Audio Manager 去访问系统的音视频服务。

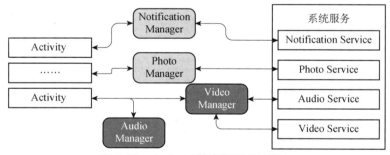

图 8-19　Android 的音视频系统服务

视频播放
（实践案例）

8.2.2　实践案例——视频播放

下面我们看一个音频的使用实例，这是一个包含"播放"、"暂停"和"停止"三个按钮的应用。单击"播放"按钮，手机播放音乐；单击"暂停"按钮，音乐暂停播放；单击"停止"按钮，音乐停止播放，如图 8-20 所示。

下面我们看一个视频的使用实例，这是一个包含"播放"、"暂停"和"重播"三个按钮的应用。单击"播放"按钮，手机播放视频；单击"暂停"按钮，视频暂停播放；单击"重播"按钮，视频重新播放，如图 8-21 所示。

图 8-20　Android 的音频实例　　　　图 8-21　Android 的视频实例

首先将播放的音频和视频文件放到 sdcard 根目录下，使用 Android 文件浏览器的上传和下载功能可以对 SD 卡根目录的文件进行上传和下载，如图 8-22 所示。

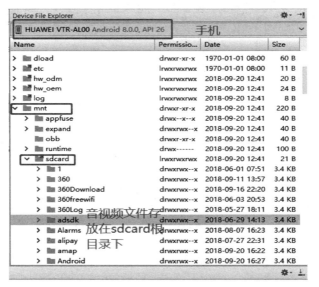

图 8-22　SD 卡根目录的文件上传和下载

Android 音频使用 MediaPlayer 类来实现。MediaPlayer 类的接口如表 8-1 所示。

表 8-1　MediaPlayer 类的接口

序号	接口名称	接口说明
1	setDataSource()	设置要播放的音频文件的路径
2	prepare()	开始播放音频前调用此方法完成准备工作
3	start()	开始或者继续播放音频
4	pause()	暂停播放音频
5	reset()	将播放器重置到初始状态

续表

序号	接口名称	接口说明
6	seekto()	从指定位置播放音频
7	stop()	停止播放音频
8	isPlaying()	判断当前是否正在播放音频

MediaPlayer 的使用步骤如图 8-23 所示。
① 创建 MediaPlayer 对象。
② 设置文件路径。
③ 利用 prePare 方法进入准备状态。
④ start 进入播放状态。
⑤ pause 进入暂停状态。
⑥ reset 停止播放。

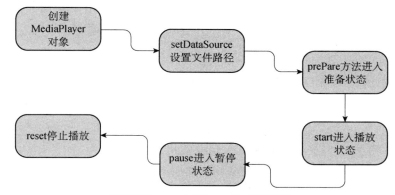

图 8-23　MediaPlayer 的使用步骤

音频文件的读写必须申请读写 SD 卡权限。
① 使用 checkSelfPermission() 检查权限；请求某个权限时要检查这个权限是否已经被用户授权，已经授权的权限重复申请可能会让用户产生厌烦；权限名称为 Manifest.permission.WRITE_EXTERNAL_STORAGE，将返回值与 PackageManager.PERMISSION_GRANTED 作比较，一致则代表有权限，不一致则代表没有权限。

使用 requestPermissions() 来申请权限，调用后系统会显示一个请求用户授权的提示对话框，App 不能配置和修改这个对话框，如果需要提示用户这个权限相关的信息或说明，需要在调用 requestPermissions() 之前处理，如图 8-24 所示。

```java
private void initData() {
    mediaPlayer = new MediaPlayer();
    if (ContextCompat.checkSelfPermission(MainActivity.this,
            Manifest.permission.WRITE_EXTERNAL_STORAGE) !=
    PackageManager.PERMISSION_GRANTED) {
        ActivityCompat.requestPermissions(MainActivity.this,
                new String[]{Manifest.permission.WRITE_EXTERNAL_STORAGE}, 1);
    } else {
        initMediaPlayer(); // 初始化 MediaPlayer
    }
}
```

图 8-24　申请读写 SD 卡权限代码

② Activity 重写 onRequestPermissionsResult()方法处理权限结果回调。当用户处理完授权操作时，系统会自动回调该方法，如果用户同意则初始化 MediaPlayer，如图 8-25 所示。

```java
@Override
public void onRequestPermissionsResult(int requestCode, String[] permissions, int[] grantResults) {
    switch (requestCode) {
        case 1:
            if (grantResults.length > 0 && grantResults[0] == PackageManager.PERMISSION_GRANTED) {
                initMediaPlayer();
            } else {
                Toast.makeText(this, "拒绝权限将无法使用程序", Toast.LENGTH_SHORT).show();
                finish();
            }
            break;
        default:
    }
}
```

图 8-25　onRequestPermissionsResult 方法代码

③ 在配置信息文件中，设置用户的 SD 卡权限，如图 8-26 所示。

```xml
<uses-permission android:name="android.permission.WRITE_EXTERNAL_STORAGE" />
```

图 8-26　配置 SD 卡权限代码

读写 SD 卡权限申请完成后再进行音频组件的初始化，如图 8-27 所示。获取文件路径，指定音频文件的路径，让 MediaPlayer 进入到准备状态。

```java
private void initMediaPlayer() {
    try {
        File file = new File(Environment.getExternalStorageDirectory(), "music.mp3");
        mediaPlayer.setDataSource(file.getPath()); // 指定音频文件的路径
        mediaPlayer.prepare(); // 让 MediaPlayer 进入到准备状态
    } catch (Exception e) {
        e.printStackTrace();
    }
}
```

图 8-27　音频组件初始化代码

主视图退出后必须销毁音频相关组件，首先让音频组件停止播放，然后释放音频组件资源，如图 8-28 所示。

```java
@Override
protected void onDestroy() {
    super.onDestroy();
    if (mediaPlayer != null) {
        mediaPlayer.stop();
        mediaPlayer.release();
    }
}
```

图 8-28　销毁音频组件代码

音频的按钮响应函数如图 8-29 所示，音频的按钮响应函数中依次完成开始播放、暂停播放和停止播放的功能。

```java
@Override
public void onClick(View v) {
    switch (v.getId()) {
        case R.id.btn_play:
            if (!mediaPlayer.isPlaying()) {
                mediaPlayer.start(); // 开始播放
            }
            break;
        case R.id.btn_pause:
            if (mediaPlayer.isPlaying()) {
                mediaPlayer.pause(); // 暂停播放
            }
            break;
        case R.id.btn_stop:
            if (mediaPlayer.isPlaying()) {
                mediaPlayer.reset(); // 停止播放
                initMediaPlayer();
            }
            break;
    }
}
```

图 8-29 音频的按钮响应函数

下面我们介绍 Android 视频的使用，Android 视频使用 VideoView 类来实现，VideoView 类的接口如表 8-2 所示。

表 8-2 VideoView 类的接口

序号	接口名称	接口说明
1	setVideoPath()	设置要播放视频文件路径
2	start()	开始或者继续播放视频
3	pause()	暂停播放视频
4	reset()	将播放器重置到初始状态
5	seekto()	从指定位置播放视频
6	stop()	停止播放视频
7	isPlaying()	判断当前是否正在播放视频
8	getDuration()	获取当前播放视频的时间长度

Android 视频使用 MediaPlayer 步骤如下所示：

① 系统申请 SD 卡写权限。
② 完成 VideoView 组件初始化。
③ 设置视频文件路径。
④ start 进入播放状态。
⑤ pause 进入暂停状态。
⑥ reset 停止播放。

视频文件的读写必须申请读写 SD 卡权限。

① 使用 checkSelfPermission()检查权限，请求某个权限前要检查这个权限是否已经被用户授权，已经授权的权限重复申请可能会让用户产生厌烦。权限名称为 Manifest.permission.WRITE_EXTERNAL_STORAGE，再将函数返回值与 PackageManager.PERMISSION_GRANTED 作比较，一致代表有权限，不一致则代表没有权限。

使用 requestPermissions() 申请权限，调用后系统会显示一个请求用户授权的提示对话框，App 不能配置和修改这个对话框，如果需要提示用户这个权限相关的信息或说明，需要在调用 requestPermissions() 之前处理，如图 8-30 所示。

```java
private void initData() {
    mediaPlayer = new MediaPlayer();
    if (ContextCompat.checkSelfPermission(MainActivity.this,
            Manifest.permission.WRITE_EXTERNAL_STORAGE) !=
    PackageManager.PERMISSION_GRANTED) {
        ActivityCompat.requestPermissions(MainActivity.this,
                new String[]{Manifest.permission.WRITE_EXTERNAL_STORAGE}, 1);
    } else {
        initVideoPath (); // 初始化 initVideoPath
    }
}
```

图 8-30　申请读写 SD 卡权限代码

② Activity 重写 onRequestPermissionsResult() 方法来处理权限结果回调。当用户处理完授权操作时，系统会自动回调该方法，如果用户同意则调用初始化方法 initVideoPath，如图 8-31 所示。

```java
@Override
public void onRequestPermissionsResult(int requestCode, String[] permissions, int[] grantResults) {
    switch (requestCode) {
        case 1:
            if (grantResults.length > 0 && grantResults[0] ==
    PackageManager.PERMISSION_GRANTED) {
                initVideoPath();
            } else {
                Toast.makeText(this, "拒绝权限将无法使用程序",
    Toast.LENGTH_SHORT).show();
                finish();
            }
            break;
        default:
    }
}
```

图 8-31　onRequestPermissionsResult 方法代码

③ 在配置信息文件中，设置用户的 SD 卡权限，如图 8-32 所示。

```xml
<uses-permission android:name="android.permission.WRITE_EXTERNAL_STORAGE" />
```

图 8-32　配置 SD 卡权限代码

读写 SD 卡权限申请完成后进行视频组件初始化，如图 8-33 所示。设置获取文件路径，指定视频文件的路径，让 mVideoView 进入到准备状态。

```java
private void initVideoPath() {
    File file = new File(Environment.getExternalStorageDirectory(), "movie.mp4");
    mVideoView.setVideoPath(file.getPath()); // 指定视频文件的路径
}
```

图 8-33　视频组件初始化代码

主视图退出时必须销毁视频相关组件，首先让视频组件停止播放，然后释放视频组件资源，如图 8-34 所示。

```
@Override
protected void onDestroy() {
    super.onDestroy();
    if (mVideoView != null) {
        mVideoView.suspend();
    }
}
```

图 8-34 销毁视频组件代码

视频的按钮响应函数如图 8-35 所示，视图的按钮响应函数中依次完成开始播放、暂停播放和重新播放的功能。

```
@Override
public void onClick(View v) {
    switch (v.getId()) {
        case R.id.btn_play:
            if (!mVideoView.isPlaying()) {
                mVideoView.start(); // 开始播放
            }
            break;
        case R.id.btn_pause:
            if (mVideoView.isPlaying()) {
                mVideoView.pause(); // 暂停播放
            }
            BroadcastReceiver
            break;
        case R.id.btn_replay:
            if (mVideoView.isPlaying()) {
                mVideoView.resume(); // 重新播放
            }
            break;
    }
}
```

图 8-35 视图的按钮响应函数

8.2.3 单元小测

判断题：

1. AudioManager 是 Android 中的一个系统服务组件。（　　）
   A. 是　　　　　　　　　　　　　　B. 否
2. VideoManager 是 Android 中的一个系统服务组件。（　　）
   A. 是　　　　　　　　　　　　　　B. 否
3. 音视频服务不需要在配置信息文件中申请权限。（　　）
   A. 是　　　　　　　　　　　　　　B. 否
4. Android 音频使用 AudioPlayer 类来实现功能。（　　）
   A. 是　　　　　　　　　　　　　　B. 否
5. Android 视频使用 MediaPlayer 类来实现功能。（　　）
   A. 是　　　　　　　　　　　　　　B. 否

选择题：

1. MediaPlayer 类中 prepare()方法的作用是（　　）。
   A. 开始或者继续播放音频
   B. 将播放器重置到初始状态
   C. 设置要播放音频文件位置
   D 开始播放音频前调用此方法完成准备工作

2. MediaPlayer 类中 start()方法的作用是（　　）。
   A. 开始或者继续播放音频
   B. 将播放器重置到初始状态
   C. 设置要播放音频文件位置
   D. 开始播放音频前调用此方法完成准备工作

3. 下面的代码完成音频组件的初始化，请补全下列代码。（　　）

```
try {
//获取文件路径
        File file = new File(Environment.getExternalStorageDirectory(), "music.mp3");
mediaPlayer.setDataSource(file.getPath()); // 指定音频文件的路径
  (   ?   )          } catch (Exception e) {
e.printStackTrace();
    }
}
```

   A. mediaPlayer.start();
   B. mediaPlayer.Playing();
   C. mediaPlayer.prepare();
   D. mediaPlayer.reset();

4. 阅读下面代码

```
private void initData() {
    mediaPlayer = new MediaPlayer();
    if (ContextCompat.checkSelfPermission(MainActivity.this, Manifest.permission.WRITE_EXTERNAL_STORAGE) != PackageManager.PERMISSION_GRANTED) {
        ActivityCompat.requestPermissions(MainActivity.this,
        new String[]{Manifest.permission.WRITE_EXTERNAL_STORAGE}, 1);
    } else {
    initMediaPlayer(); // 初始化 MediaPlayer
    }
};
```

代码中 "1" 的作用是什么？（　　）
   A. 用户申请权限的标志码
   B. 用户申请权限的校验码
   C. 用户申请权限的标志码
   D. 无任何作用

5. ViewVideo 类中 start()方法的作用是（　　）。
   A. 设置要播放视频文件路径
   B. 开始或者继续播放视频
   C. 将播放器重置到初始状态
   D. 开始播放视频频前调用此方法完成准备工作

6. ViewVideo 类中 reset()方法的作用是（　　）。
   A. 设置要播放视频文件路径

B. 开始或者继续播放视频
C. 将播放器重置到初始状态
D. 开始播放视频前调用此方法完成准备工作

7. 下面实现视频组件初始化功能，请补全代码。（　　）

```
private void initVideoPath() {
//获取视频文件路径
  File file = new File(Environment.getExternalStorageDirectory(), "movie.mp4");
  (    ?    )
}
```

A. mVideoView.setPath(file.getPath());
B. mVideoView.setPathDirectory(file.getPath());
C. mVideoView.setVideoPath(file.getPath());
D. mVideoView.setVideoPathDirectory(file.getPath());

本章课后练习和程序源代码

第 8 章源代码及课后习题

9 网络服务

Android 的网络服务、Android 的网络框架、Android 的 JSON 协议、Android 的 Volley 框架。

1. 熟练应用 Android 的 WebView 服务。
2. 熟练使用 Android 的 OkHttp 协议访问网络。
3. 熟练使用网络 JSON 协议进行网络服务。
4. 熟练使用网络 Volley 框架解析网络数据。

 ## 9.1 网络服务概述

9.1.1 知识点讲解——网络服务

网络服务概述(慕课)

本节我们介绍 Android 的网络服务,现在很多 App 里都内置了 Web 网页(Hyprid App),比如很多电商平台,淘宝、京东、聚划算等。那么这种服务该如何实现呢?其实这是通过 Android 里一个叫 WebView 的组件来实现的。Android 手机中内置了一个高性能的 Webkit 内核浏览器,在 SDK 中被打包成 WebView 组件;这个组件主要应用于网页的加载,可以直接显示网页,也可以加载 HTML 文件,实现复杂的布局界面。

下面介绍一个使用 WebView 浏览网页的例子,如图 9-1 所示,内置了新浪的 Web 网页,网页和普通网页一样,可以支持页面的跳转和内部访问。

图 9-1　使用 WebView 浏览网页

下面首先看一下 WebView 布局的实现，如图 9-2 所示。在线性布局中，新增一个 WebView 控件，id 设置为"web_view"，宽度和高度适配整个屏幕。

```xml
<?xml version="1.0" encoding="utf-8"?>
<LinearLayout xmlns:android="http://schemas.android.com/apk/res/android"
    android:layout_width="match_parent"
    android:layout_height="match_parent" >
    <WebView
        android:id="@+id/web_view"
        android:layout_width="match_parent"
        android:layout_height="match_parent" />
</LinearLayout>
```

图 9-2　WebView 布局

WebView 的使用比较简单，我们看一下常用的用法，如图 9-3 所示。WebView 组件是 Android 的一个 Web 访问组件，使用 findViewById 对 WebView 控件进行初始化；使用 setJavaScriptEnabled 设置 WebView 控件支持 JavaScript 脚本；使用 setWebViewClient 设置网络客户端；使用 loadUrl 设置网页地址。

```java
package com.example.webviewtest;
import android.os.Bundle;
import android.support.v7.app.AppCompatActivity;
import android.webkit.WebView;
import android.webkit.WebViewClient;

public class MainActivity extends AppCompatActivity {
    private WebView mWebView;
    @Override
    protected void onCreate(Bundle savedInstanceState) {
        super.onCreate(savedInstanceState);
        setContentView(R.layout.activity_main);
        initView();
        initData();
    }
    private void initData() {
        mWebView.getSettings().setJavaScriptEnabled(true);// 支持 JavaScript 脚本
        mWebView.setWebViewClient(new WebViewClient());// 设置网络客户端
        mWebView.loadUrl("http://www.sina.com.cn");// 设置网页地址
```

图 9-3　视图的按钮响应函数

```
    }
    private void initView() {
        mWebView = (WebView) findViewById(R.id.web_view);
    }
}
```

图 9-3　视图的按钮响应函数（续）

在 Android 的信息配置文件 AndroidManifest.xml 中增加网络访问权限"android.permission.INTERNET"，如图 9-4 所示。

```xml
<?xml version="1.0" encoding="utf-8"?>
<manifest xmlns:android="http://schemas.android.com/apk/res/android"
    package="com.example.webviewtest">
    <uses-permission android:name="android.permission.INTERNET" />
    <application
        android:allowBackup="true"
        android:icon="@mipmap/ic_launcher"
        android:label="@string/app_name"
        android:supportsRtl="true"
        android:theme="@style/AppTheme">
        <activity android:name=".MainActivity">
            <intent-filter>
                <action android:name="android.intent.action.MAIN" />

                <category android:name="android.intent.category.LAUNCHER" />
            </intent-filter>
        </activity>
    </application>
</manifest>
```

图 9-4　AndroidManifest.xml

WebView 方法适合访问简单的网页，现在一般都将 HTML 放置在 Tomcat 服务器上，用服务器显示 HTML 文件，客户端获取并显示内容，如图 9-5 所示。

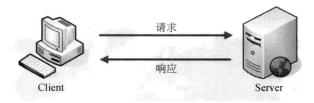

图 9-5　C/S 模式

说到 HTTP 协议，那必须要先了解 WWW。WWW 是环球信息网（World Wide Web）的缩写，也可以简称为 Web，中文名字为"万维网"。简单来说，WWW 是以 Internet 作为传输媒介的一个应用系统，WWW 上基本的传输单位是 Web 网页。WWW 的工作是基于 B/S 模型的，由 Web 浏览器和 Web 服务器构成，两者之间采用超文本传输协议 HTTP 进行通信。

HTTP 协议是基于 TCP/IP 协议之上的协议，是 Web 浏览器和 Web 服务器之间的应用层的协议，是通用的、无状态的面向对象的协议。

数据在 Internet 上传输，一般通过三种网络协议来实现信息的发送和接收。

◆ HTTP 协议：最常用的协议，是建立在 TCP/IP 基础上的。
◆ FTP 协议：文件传输协议。
◆ TCP/IP 协议：它也是底层的协议，其他的方式必须通过它，但是想要实现这种协议必须实现 Socket 编程，这种方法常用来进行上传一些比较大的文件、视频，需要断点续传。

9.1.2　实践案例——使用 HTTP 协议访问网络

WebView 和 Http 服务
（实践案例）

下面我们看一个使用 HTTP 协议访问网络的实例，如图 9-6 所示；输入网络地址 https://www.sohu.com，如果访问成功，则将网络返回的 HTML 文件的内容全部显示到文本框中；输入一个地址 https://192.168.1.100，访问不成功后提示错误信息。

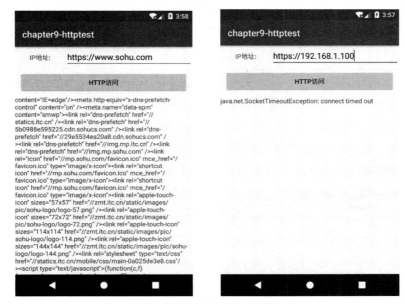

图 9-6　使用 HTTP 协议访问实例

如图 9-7 所示的是一种 HTTP 协议访问的实现架构，网络异步线程与主线程通过 EventBus 通信；主视图中启动网络异步线程，自定义线程中包含了网络请求成功和网络请求失败两种可能；网络请求的结果通过 EventBus 将网络请求的数据发送到主视图。采用这种架构的数据结构比较统一，但缺点是 EventBus 并不能处理复杂的网络数据。

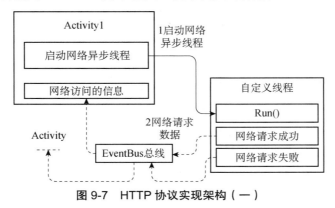

图 9-7　HTTP 协议实现架构（一）

如图 9-8 所示的是另外一种 HTTP 协议访问的实现架构，异步线程与主线程通过接口进行通信；主视图中启动网络异步线程，同时在网络异步线程中设置了一个回调的接口 HttpListenser。HttpListenser 包含两个接口，其中 onSucess 处理网络成功访问的消息，onFailed 处理网络访问失败的消息。网络异步线程处理网络请求成功后，通过 onSucess 和 onFailed 将网络请求的数据发送到主视图。这种架构可以处理复杂的网络数据并且网络处理效率高，但是必须实现接口，不同的网络处理请求需要实现不同的接口。

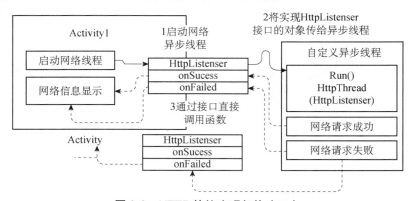

图 9-8　HTTP 协议实现架构（二）

首先我们实现 HTTP 实例的界面布局，如图 9-9 所示。

```xml
<?xml version="1.0" encoding="utf-8"?>
<LinearLayout xmlns:android="http://schemas.android.com/apk/res/android"
    xmlns:app="http://schemas.android.com/apk/res-auto"
    xmlns:tools="http://schemas.android.com/tools"
    android:layout_width="match_parent"
    android:layout_height="match_parent"
    android:orientation="vertical"
    tools:context=".MainActivity">
    <LinearLayout
        android:layout_width="match_parent"
        android:layout_height="wrap_content"
        android:orientation="horizontal">
        <TextView
            android:id="@+id/textView"
            android:layout_width="wrap_content"
            android:layout_height="wrap_content"
            android:layout_weight="1"
            android:gravity="center"
            android:text="IP 地址:" />
        <EditText
            android:id="@+id/edt_ipadress"
            android:layout_width="wrap_content"
            android:layout_height="wrap_content"
            android:layout_weight="1"
            android:ems="10"
            android:hint="192.168.1.2"
            android:inputType="textPersonName"
            android:text="192.168.1.100" />
    </LinearLayout>

    <Button
        android:id="@+id/btn_http"
        android:layout_width="match_parent"
        android:layout_height="wrap_content"
        android:layout_margin="10dp"
        android:text="Http 访问" />
```

图 9-9　HTTP 实例的界面布局

```xml
<ScrollView
    android:layout_width="match_parent"
    android:layout_height="match_parent">
    <LinearLayout
        android:layout_width="match_parent"
        android:layout_height="wrap_content"
        android:orientation="vertical" >
        <TextView
            android:id="@+id/textView_content"
            android:layout_width="match_parent"
            android:layout_height="wrap_content"
            android:text="TextView" />
    </LinearLayout>
</ScrollView>
</LinearLayout>
```

图 9-9　HTTP 实例的界面布局（续）

HTTP 实例的界面布局的效果如图 9-10 所示。

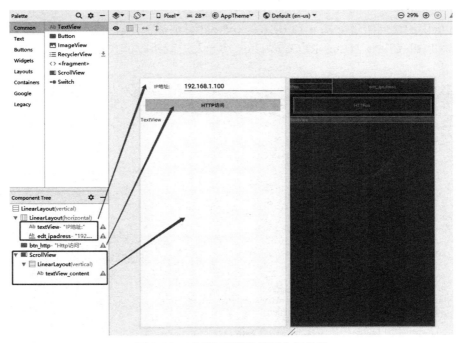

图 9-10　HTTP 实例的界面布局效果

下面介绍一下 HTTP 协议的实现流程。

① 监听接口设计，新建一个 HttpListener 的监听接口，包含 onSuccess 和 onFailed 两个接口，如图 9-11 所示。

```java
package cn.edu.sziit.chapter9_httptest;
public interface HttpListener {
    void onSuccess(final String strResponse);
    void onFailed(final String strResponse);
}
```

图 9-11　HttpListener 的监听接口

② 新建一个 Http 线程类，如图 9-12 所示，新建一个 mHttpListener 变量和 strUrl 变量，

并根据两个变量生成构造函数。

```java
package cn.edu.sziit.chapter9_httptest;
import java.io.BufferedReader;
import java.io.IOException;
import java.io.InputStream;
import java.io.InputStreamReader;
import java.net.HttpURLConnection;
import java.net.URL;
public class HttpThread extends Thread {
    private HttpListener mHttpListener;
    private String strUrl;
    public HttpThread(String strUrl,HttpListener mHttpListener)
    {
        this.strUrl=strUrl;
        this.mHttpListener=mHttpListener;
    }
}
```

图 9-12　Http 线程类

HTTP 协议的访问实现如图 9-13 所示；Android 目前提供两种 HTTP 通信方式：HttpURLConnection 和 HttpClient。HttpURLConnection 用于发送或接收流式数据，因此比较适合上传/下载文件。

```java
public class HttpThread extends Thread {
@Override
public void run() {
    super.run();
    //打开http 链接
    HttpURLConnection mHttpURLConnection = null;
    BufferedReader mBufferedReader = null;
    InputStream in=null;
    try {
        URL mUrl = new URL(strUrl);
        mHttpURLConnection = (HttpURLConnection) mUrl.openConnection();
        //设置超时时间
        mHttpURLConnection.setConnectTimeout(5000);
        //指定请求方式为GET 方式
        mHttpURLConnection.setRequestMethod("GET");
        mHttpURLConnection.setReadTimeout(5000);
        //不用再去判断状态码是否为200
        in= mHttpURLConnection.getInputStream();
        mBufferedReader = new BufferedReader(new InputStreamReader(in));
        StringBuilder response = new StringBuilder();
        String line;
        while ((line = mBufferedReader.readLine()) != null) {
            response.append(line);
        }
        mHttpListener.onSuccess(response.toString());
    } catch (Exception e) {
        // TODO: handle exception
        e.printStackTrace();
        mHttpListener.onFailed(e.toString());
    }finally {
        if (mBufferedReader!=null)
        {
            try {
                mBufferedReader.close();
                in.close();
            }catch (IOException e)
            {
                e.printStackTrace();
            }
```

图 9-13　HTTP 协议的访问实现

```
        if(mHttpURLConnection!=null)
        {
            mHttpURLConnection.disconnect();
        }
    }
}
```

图 9-13　HTTP 协议的访问实现（续）

首先我们进行网络设置，主要流程介绍如下：
① 新建 HttpURLConnection 访问对象。
② 新建 BufferedReader 缓存。
③ 新建 InputStream 输入流，由于要进行文件输入输出流操作，网络请求代码需要使用 try catch 来作保护。
④ 新建 URL 对象。
⑤ 根据 URL 初始化 HttpURLConnection 访问对象。
⑥ 设置网络访问超时时间为 5 秒。
⑦ 指定请求方式为 GET 方式。
⑧ 设置读取数据超时时间为 5 秒。

HTTP 协议返回数据后的处理，主要流程如下：
① 获取输入流。
② 将输入流放入缓存中。
③ 新建字符串缓存。
④ 按行读取输入流缓存并将网络请求所有内容存入字符串缓存中。
⑤ 网络请求成功将信息提交回调接口。
⑥ 网络请求失败将错误信息提交回调接口。

网络访问完成后需要回收资源，流程如下：
① 关闭缓存。
② 关闭输入流。
③ 关闭网络连接。

主视图中实现主界面的监听接口，由于网络请求后的数据最后都是在主视图中显示的，因此监听接口需要在主视图中实现，如图 9-14 所示。
① 新建网络监听器对象。
② 初始化网络监听器对象。
③ 实现 onSuccess 接口，由于网络访问后的回调属于异步线程，需要使用 runOnUiThread 完成对组件的访问。
④ 网络访问成功后获取数据显示。
⑤ 实现 onFailed 接口，由于网络访问成功后回调属于异步线程，需要使用 runOnUiThread 来完成对组件的访问。
⑥ 网络访问失败后获取数据显示。

```java
package cn.edu.sziit.chapter9_httptest;
import android.os.Bundle;
import android.support.v7.app.AppCompatActivity;
import android.text.TextUtils;
import android.view.View;
import android.widget.Button;
import android.widget.EditText;
import android.widget.TextView;
import android.widget.Toast;

public class MainActivity extends AppCompatActivity implements View.OnClickListener {
    private HttpListener mHttpListener;
    private TextView mTextView;
    private EditText mEdtIpadress;
    private Button mBtnHttp;
    private TextView mTextViewContent;
    @Override
    protected void onCreate(Bundle savedInstanceState) {
        super.onCreate(savedInstanceState);
        setContentView(R.layout.activity_main);
        initView();
        initData();
    }
    private void initData() {
        mHttpListener = new HttpListener() {
            @Override
            public void onSuccess(final String strResponse) {
                runOnUiThread(new Runnable() {
                    @Override
                    public void run() {
                        //主线程进行 UI 操作
                        mTextViewContent.setText(strResponse);
                    }
                });
            }
            @Override
            public void onFailed(final String strResponse) {
                runOnUiThread(new Runnable() {
                    @Override
                    public void run() {
                        //主线程进行 UI 操作
                        mTextViewContent.setText(strResponse);
                    }
                });

            }
        };
    }
    private void initView() {
        mTextView = (TextView) findViewById(R.id.textView);
        mEdtIpadress = (EditText) findViewById(R.id.edt_ipadress);
        mBtnHttp = (Button) findViewById(R.id.btn_http);

        mBtnHttp.setOnClickListener(this);
        mTextViewContent = (TextView) findViewById(R.id.textView_content);
        mTextViewContent.setOnClickListener(this);
    }
    @Override
    public void onClick(View v) {
        switch (v.getId()) {
            case R.id.btn_http:
                submit();
                break;
        }
    }
}
```

图 9-14　主界面的监听接口

在主视图中启动 HTTP 线程，如图 9-15 所示。

```java
package cn.edu.sziit.chapter9_httptest;
import android.os.Bundle;
import android.support.v7.app.AppCompatActivity;
import android.text.TextUtils;
import android.view.View;
import android.widget.Button;
import android.widget.EditText;
import android.widget.TextView;
import android.widget.Toast;
public class MainActivity extends AppCompatActivity implements View.OnClickListener {
    private void submit() {
        // validate
        String ipadress = mEdtIpadress.getText().toString().trim();
        if (TextUtils.isEmpty(ipadress)) {
            Toast.makeText(this, "ip 地址不能为空", Toast.LENGTH_SHORT).show();
            return;
        }
        mTextViewContent.setText("");
        HttpThread mHttpThread=new HttpThread(ipadress,mHttpListener);
        mHttpThread.start();
    }
}
```

图 9-15　启动 HTTP 线程

① 新建 HttpThread 线程对象并使用 ipadress 和 mHttpListener 监听器对象初始化构造函数。
② 使用 mHttpThread.start()方法启动线程。

增加网络访问权限，配置信息文件中 Android 的网络权限为"android.permission. INTERNET"，如图 9-16 所示。

```xml
<?xml version="1.0" encoding="utf-8"?>
<manifest xmlns:android="http://schemas.android.com/apk/res/android"
    package="cn.edu.sziit.chapter9_httptest">
    <uses-permission android:name="android.permission.INTERNET" />
    <application
        android:allowBackup="true"
        android:icon="@mipmap/ic_launcher"
        android:label="@string/app_name"
        android:roundIcon="@mipmap/ic_launcher_round"
        android:supportsRtl="true"
        android:theme="@style/AppTheme">
        <activity android:name=".MainActivity">
            <intent-filter>
                <action android:name="android.intent.action.MAIN" />

                <category android:name="android.intent.category.LAUNCHER" />
            </intent-filter>
        </activity>
    </application>
</manifest>
```

图 9-16　配置 Android 的网络权限

9.1.3　单元小测

判断题：

1. WebView 是 Android 中的一个系统组件，用于网络访问。（　　）

A. 是　　　　　　　　　　　　B. 否
2. 网络服务不需要在配置信息文件中申请权限。（　　）
A. 是　　　　　　　　　　　　B. 否
3. HttpURLConnection 是一个标准的 Java 类。（　　）
A. 是　　　　　　　　　　　　B. 否
4. HttpURLConnection 访问网络，不需要创建 HttpURLConnection 对象。（　　）
A. 是　　　　　　　　　　　　B. 否

选择题：

1. 在配置文件中申请网络的权限，下面哪一个是正确的？（　　）

A. uses-permission android:name="android.permission.NET"

B. uses-permission android:name="android.permission.NETWORK"

C. uses-permission android:name="android.permission.INTERNET.GET"

D. uses-permission android:name="android.permission.INTERNET"

2. 下面代码用于在 WebView 组件中访问网页，请补全代码。（　　）

```
private void initData() {
    mWebView.getSettings().setJavaScriptEnabled(true); //支持 JavaScript 脚本
    mWebView.setWebViewClient(new WebViewClient()); //设置网络客户端
    （ ? ）
}
```

A. mWebView.loadHttp("http://www.sina.com.cn")

B. mWebView.loadHttpAddres("http://www.sina.com.cn")

C. mWebView.loadUrl("http://www.sina.com.cn")

D. mWebView.loadUrlAddres("http://www.sina.com.cn")

3. 下列关于 HttpURLConnection 访问网络的用法中，说法错误的是（　　）。

A. HttpURLConnection 对象需要设置请求网络的超时时间

B. HttpURLConnection 对象需要设置请求网络的方式

C. 需要通过 new 关键字创建 HttpURLConnection 对象

D. 访问网络不需要关闭网络连接

4. 下面是 HttpURLConnection 设置的代码，请补全代码。（　　）

```
{
URL mUrl = new URL(strUrl); //新建 URL 对象
mHttpURLConnection = (HttpURLConnection)
    mUrl.openConnection();
mHttpURLConnection.setConnectTimeout(5000);
mHttpURLConnection.setRequestMethod("GET");
//设置读取数据超时时间  （ ? ）}
```

A. mHttpURLConnection.setDataTimeout(5000);

B. mHttpURLConnection.setGetDataTimeout(5000);

C. mHttpURLConnection.setReadDataTimeout(5000);

D. mHttpURLConnection.setReadTimeout(5000);

9.2 网络框架

9.2.1 知识点讲解——网络框架

本节我们以 OkHttp 为例介绍网络框架。图 9-17 所示的是使用 OkHttp 框架访问网络的一个实例，输入网络地址 https：//www.sohu.com，如果访问成功，则将网络返回的 HTML 文件的内容全部显示到文本框中，访问一个地址 https://192.168.1.100，访问不成功后提示错误信息。

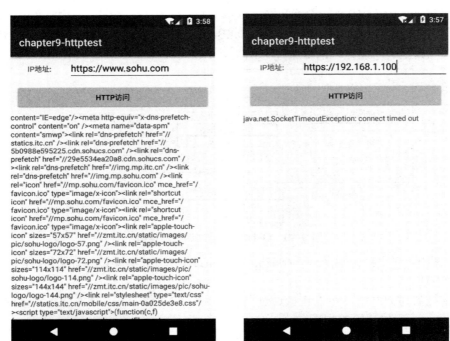

图 9-17　OkHttp 实例

OkHttp 框架是一个处理网络请求的开源项目，是 Android 端最火热的轻量级框架，主要的特点如下：

- ◆ 允许连接到同一个主机地址的所有请求，提高请求效率。
- ◆ 共享 Socket，减少对服务器的请求次数。
- ◆ 通过连接池，减少了请求延迟。
- ◆ 通过缓存响应数据来减少重复的网络请求。
- ◆ 减少了对数据流量的消耗。

OkHttp 框架使用 HttpURLConnection 进行通信，对 HttpURLConnection 进行封装，调用网络请求比较方便，OkHttp 的调用框架如图 9-18 所示。OkHttp 框架封装了 HttpURLConnection 请求部分，整体的代码架构比较简洁。

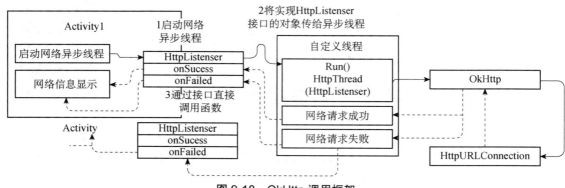

图 9-18　OkHttp 调用框架

HttpURLConnection 的连接设置和网络读取数据都是比较复杂的，使用 OkHttp 框架，网络访问代码简洁轻便。OkHttp 框架实现了 CallBack 异步网络回调接口，可以使用 OkHttp 的 CallBack 进行数据的回调处理的框架，如图 9-19 所示。

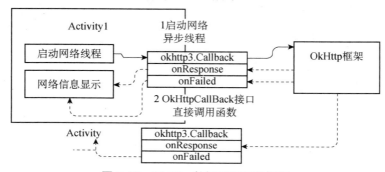

图 9-19　OkHttp 数据回调处理框架

在主视图中可以直接初始化一个 OkHttp 的 CallBack 对象，实现 CallBack 接口的 onResponse 和 onFailed 方法。

OkHttp 框架访问网络后可以直接通过调用 onResponse 和 onFailed 方法将网络请求的数据传送给主视图。

OkHttp 框架
（实践案例）

9.2.2　实践案例——网络框架

要使用 OkHttp 框架，可以直接在 build.grade 的 dependencies 中增加实现 OkHttp 的路径。OkHttp 开源地址为"com.squareup.okhttp3:okhttp:3.11.0"，build.gradle 文件实现如图 9-20 所示。

```
apply plugin: 'com.android.application'
android {
    compileSdkVersion 27
    defaultConfig {
        applicationId "cn.edu.sziit.chapter9_okhttptest"
        minSdkVersion 15
        targetSdkVersion 27
        versionCode 1
        versionName "1.0"
        testInstrumentationRunner
"android.support.test.runner.AndroidJUnitRunner"
```

图 9-20　build.gradle 文件实现

```
        }
        buildTypes {
            release {
                minifyEnabled false
                proguardFiles getDefaultProguardFile('proguard-android.txt'),
'proguard-rules.pro'
            }
        }
    }
dependencies {
        implementation fileTree(dir: 'libs', include: ['*.jar'])
        implementation 'com.android.support:appcompat-v7:27.1.1'
        implementation 'com.android.support.constraint:constraint-layout:1.1.3'
        testImplementation 'junit:junit:4.12'
        implementation 'com.squareup.okhttp3:okhttp:3.11.0'
        implementation 'com.google.code.gson:gson:2.8.5'
        androidTestImplementation 'com.android.support.test:runner:1.0.2'
        androidTestImplementation
'com.android.support.test.espresso:espresso-core:3.0.2'
    }
```

图 9-20　build.gradle 文件实现（续）

增加网络访问权限，配置信息文件中 Android 的网络权限为"android.permission. INTERNET"，如图 9-21 所示。

```
<?xml version="1.0" encoding="utf-8"?>
<manifest xmlns:android="http://schemas.android.com/apk/res/android"
    package="cn.edu.sziit.chapter9_httptest">
    <uses-permission android:name="android.permission.INTERNET" />
    <application
        android:allowBackup="true"
        android:icon="@mipmap/ic_launcher"
        android:label="@string/app_name"
        android:roundIcon="@mipmap/ic_launcher_round"
        android:supportsRtl="true"
        android:theme="@style/AppTheme">
        <activity android:name=".MainActivity">
            <intent-filter>
                <action android:name="android.intent.action.MAIN" />

                <category android:name="android.intent.category.LAUNCHER" />
            </intent-filter>
        </activity>
    </application>
</manifest>
```

图 9-21　配置 Android 的网络权限

主视图中实现主界面的监听接口，由于网络请求后的数据最后都是在主视图中显示的，因此监听接口需要在主视图中实现，如图 9-22 所示。

① 新建 okhttp3.Callback 网络监听器对象。
② 初始化网络监听器对象。
③ 实现 onFailure 接口。
④ 网络访问失败后，获取数据显示。
⑤ 实现 onResponse 接口。
⑥ 网络访问成功后，获取数据显示。

```java
package cn.edu.sziit.chapter9_okhttptest;
import android.os.Bundle;
import android.support.v7.app.AppCompatActivity;
import android.text.TextUtils;
import android.view.View;
import android.widget.Button;
import android.widget.EditText;
import android.widget.TextView;
import android.widget.Toast;
import java.io.IOException;
import okhttp3.Call;
import okhttp3.OkHttpClient;
import okhttp3.Request;
import okhttp3.Response;
public class MainActivity extends AppCompatActivity implements View.OnClickListener {
    private HttpListener mHttpListener;
    private TextView mTextView;
    private EditText mEdtIpadress;
    private Button mBtnHttp;
    private TextView mTextViewContent;
    private okhttp3.Callback mokhttp3Callback;
    @Override
    protected void onCreate(Bundle savedInstanceState) {
        super.onCreate(savedInstanceState);
        setContentView(R.layout.activity_main);
        initView();
        initData();
    }
    private void initData() {
        mokhttp3Callback=new okhttp3.Callback(){
            @Override
            public void onFailure(Call call, IOException e) {
                mTextViewContent.setText(e.toString());
            }
            @Override
            public void onResponse(Call call, final Response response) throws IOException {
                runOnUiThread(new Runnable() {
                    @Override
                    public void run() {
                        //主线程进行UI 操作
                        mTextViewContent.setText(response.body().toString());
                    }
                });
            }
        };
    }
    private void initView() {
        mTextView = (TextView) findViewById(R.id.textView);
        mEdtIpadress = (EditText) findViewById(R.id.edt_ipadress);
        mBtnHttp = (Button) findViewById(R.id.btn_http);

        mBtnHttp.setOnClickListener(this);
        mTextViewContent = (TextView) findViewById(R.id.textView_content);
        mTextViewContent.setOnClickListener(this);
    }
    @Override
    public void onClick(View v) {
        switch (v.getId()) {
            case R.id.btn_http:
                submit();
                break;
        }
    }
}
```

图 9-22　主视图实现监听接口

主视图中启动 OkHttp 网络请求访问，如图 9-23 所示。
① 创建 OkHttpClienet 实例。

```
package cn.edu.sziit.chapter9_okhttptest;
public class MainActivity extends AppCompatActivity implements View.OnClickListener {
    private void submit() {
        // validate
        String ipadress = mEdtIpadress.getText().toString().trim();
        if (TextUtils.isEmpty(ipadress)) {
            Toast.makeText(this, "ip 地址不能为空", Toast.LENGTH_SHORT).show();
            return;
        }
        mTextViewContent.setText("");
        OkHttpClient mOkHttpClient=new OkHttpClient();
        Request mRequest=new Request.Builder().url(ipadress).build();
        mOkHttpClient.newCall(mRequest).enqueue(mokhttp3Callback);
    }
}
```

图 9-23　启动 OkHttp 网络请求

② 根据网络请求的 IP 地址创建 Request 请求对象。

③ 使用 newCall 创建一个 Call 对象，调用 execute 发送请求获取数据，并设置回调接口为 okhttp3.Callback 监听器对象。

上面就是使用 OkHttp 网络框架访问网络的全部过程，可以看出，使用 OkHttp 网络框架，网络访问设置、网络请求和网络返回处理都非常简便。

9.2.3　单元小测

判断题：

1. OkHttp 是一个网络处理框架，是对 URLConnection 的封装。（　　）
A. 是　　　　　　　　　　　　　　　B. 否
2. OkHttp 已对网络进行了封装，使用 OkHttp 不需要在配置信息文件中申请权限。（　　）
A. 是　　　　　　　　　　　　　　　B. 否
3. OkHttp 是一个 Android 的原生态的网络处理框架，不需要导入第三方类。（　　）
A. 是　　　　　　　　　　　　　　　B. 否
4. OkHttpClient 访问网络不需要创建 OkHttpClient 对象。（　　）
A. 是　　　　　　　　　　　　　　　B. 否

选择题：

1. OkHttp 框架的优点主要包括（　　）。
A. 减少对服务器的请求次数　　　　　B. 减少了请求延迟
C. 减少重复的网络请求　　　　　　　D. 提高请求效率
2. okhttp3.Callback 类的主要作用是（　　）。
A. 用于网络的请求　　　　　　　　　B. 用于网络的数据访问
C. 用于请求网络数据的回调处理　　　D. 用于网络服务的反馈
3. okhttp3.Callback 类必须实现的两个接口是（　　）。
A. onFail()　　　B. onFailure()　　　C. onResponse()　　　D. onSuccess()
4. 使用 OkHttp 网络请求访问，请补全代码。（　　）
mOkHttpClient=new OkHttpClient(); //创建 OkHttpClienet 实例

Request mRequest=new Request.Builder().url(ipadress).build(); //创建 Request 请求对象
(？)

A. mOkHttpClient.newCall(mRequest).execute(mokhttp3Callback);
B. mOkHttpClient.newCall(mRequest).enqueue(mokhttp3Callback);
C. mOkHttpClient.newCall(mRequest).loadUrl(mokhttp3Callback);
D. mOkHttpClient.newCall(mRequest).inqueue(mokhttp3Callback);

9.3 JSON 协议

9.3.1 知识点讲解——JSON 协议

JSON 协议
（慕课）

本节我们介绍 JSON 协议。JSON 是一种轻量级的数据交换格式，简洁和清晰的层次结构使得其成为理想的数据交换语言。JSON 易于机器解析和生成，可以有效地提升网络传输效率。JSON 采用完全独立于编程语言的文本格式来存储和表示数据。

JSON 的特点是便于跨平台使用，广泛应用于服务器和各种客户端的数据交换。如图 9-24 所示的是 JSON 的使用场景。

图 9-24　JSON 使用场景

下面我们看一个具体的 JSON 例子，如图 9-25 所示。
◆ 任何支持类型都可以使用 JSON 表示，比如字符串、数字、对象、数组。
◆ JSON 数据结构为{ "key1"："value1"，"key2"："value2"，...} 的键值对结构。
◆ JSON 也可以表示数组，数组为[{{ "key1"："value1"，"key2"："value2"，…},…]的键值对结构。

```
{   "ERRMSG": "成功",   "WCurrent": 18,
"ROWS_DETAIL": [
{"temperature": "22~30",      "WData": "2018-10-24"    },
{ "temperature": "22~25",     "WData": "2018-10-25"    },
{"temperature": "21~29",      "WData": "2018-10-26"    },
{"temperature": "20~23",      "WData": "2018-10-27"    },
{"temperature": "22~29",      "WData": "2018-10-28"    },
{"temperature": "20~28",      "WData": "2018-10-29"    } ],
"RESULT": "S " }
```

图 9-25　JSON 语法格式

JSON 数据访问主要需要如下三方面的设置：

- 服务器地址 URL，比如以下就是获取天气信息的 URL 地址，http://139.199.220.137:8080/api/v2/get_weather。
- 客户端向服务器传递参数有 POST 和 GET 两种方式。
- GET 将参数封装到 URL 中，在网络上进行"透明"传播；POST 将参数封装到 Request 的 Body 中，在网络上进行"不透明"传播。

JSON 数据访问服务器接口如表 9-1 所示。

表 9-1　JSON 数据访问服务器接口

功能说明	气象信息查询			
调用路径	http://139.199.220.137:8080/api/v2/get_weather			
HTTP 请求	POST			
入参	■单表　□多表　■单值　□多值			
参数名	类型	描述	必需	
UserName	varchar	账号	■	
返回值	■单表　□多表　■单值　□多值			
参数名	类型	描述	必需	
RUSULT	varchar	成功 S，失败 F	■	
ERRMSG	varchar	提示信息	■	
WCurrent	varchar	当前温度	■	
ROWS_DETAIL	varchar	数组	■多值	
参数名	类型	描述	必需	
WData	varchar	当前温度	■	
temperature	varchar	当前温度	■	

- 调用路径：一般由服务器提供。
- HTTP 请求：分为 GET 和 POST 两种方式。
- 入参：如果采用的是 POST 方式，需要将参数封装到 Request 的 Body 中。
- 返回值：服务器返回的是 JSON 数据格式，客户端根据格式进行解析和提取。

JSON 数据访问接口样例如表 9-2 所示。

表 9-2　JSON 数据访问接口样例

功能说明	气象信息查询访问样例
JSON 参数	{"UserName":"user1"}
服务器返回	{ "ERRMSG": "成功", "WCurrent": 18, "ROWS_DETAIL": [　{"temperature": "22~30",　　　　"WData": "2018-10-24"　　}, 　{ "temperature": "22~25",　　　　"WData": "2018-10-25"　　}, 　{"temperature": "21~29",　　　　"WData": "2018-10-26"　　}, 　{"temperature": "20~23",　　　　"WData": "2018-10-27"　　}, 　{"temperature": "22~29",　　　　"WData": "2018-10-28"　　}, 　{"temperature": "20~28",　　　　"WData": "2018-10-29"　　}], "RESULT": "S "}
JSON 访问 代码（自动生成）	OkHttpClient client = new OkHttpClient(); MediaType mediaType = MediaType.parse("application/json"); RequestBody body = RequestBody.create(mediaType, "{\"UserName\":\"user1\"}"); Request request = new Request.Builder() 　.url("http://139.199.220.137:8080/api/v2/get_weather") 　.post(body).build(); Response response = client.newCall(request).execute();

- ◆ JSON 参数：客户端请求访问需要将参数按照 JSON 格式封装到 Request 的 Body 中。
- ◆ 返回值，服务器返回的 JSON 数据格式，客户端根据格式进行解析和提取。
- ◆ JSON 访问代码可以由工具自动生成，大大降低了网络访问的编码难度。

开发人员还需要调试一个接口是否运行正常，服务器是否能够正确处理各种 HTTP 请求。

在实际的开发中，用户的大部分数据都需要通过 HTTP 请求来与服务器进行交互。postman 工具非常方便用于接口的测试，JSON 数据接口测试可以使用 postman 工具，如图 9-26 所示。新建一个接口文件夹并命名为"json 数据讲解"，新建接口并命名为"2.10.1 气象信息查询"后保存。

图 9-26　新建 JSON 测试接口

postman 工具中需要对接口进行设置和测试，如图 9-27 所示的是测试流程。

① 选择 POST 协议。
② 输入服务器的 URL 参数。
③ 将 POST 参数按照 JSON 格式封装到 Request 的 Body 中。
④ 单击"Send 发送"按钮。
⑤ 服务器返回 JSON 数据。

图 9-27　接口测试

单击"Code"按钮，选择自动生成 Java 访问网络服务器的代码，选择 Java 代码后可以自动生成 OkHttp 框架访问代码，如图 9-28 所示。

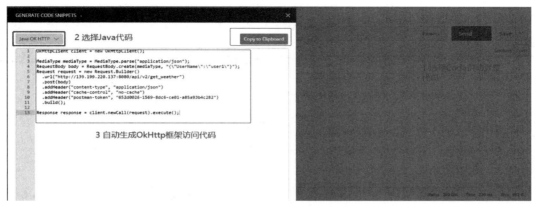

图 9-28 自动生成代码

下面介绍 GsonFormat 插件。GsonFormat 主要用于使用 Gson 库将 JSONObject 格式的 String 解析成实体。安装过程如图 9-29 所示。在 File-Setting-Plugins 中进入插件窗口，搜索 GsonFormat 后，单击"下载"按钮并安装。

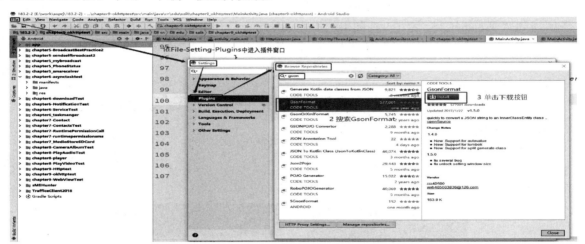

图 9-29 GsonFormat 插件的下载和安装

GsonFormat 插件可以直接将服务器返回的 JSON 数据格式转换为程序可以直接使用的数据类对象，如图 9-30 所示。

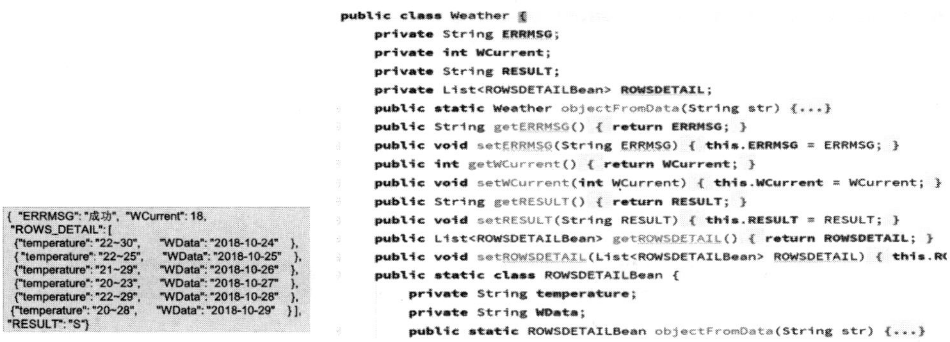

图 9-30 GsonFormat 插件的使用

GsonFormat 插件自动生成 JSON 数据 Bean 类的步骤如图 9-31 所示。

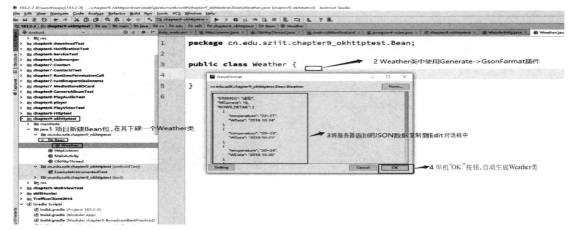

图 9-31　GsonFormat 自动生成类

① 项目中新建 Bean 包，在 Bean 包下新建一个 Weather 类。
② 进入 Weather 类，右击，在弹出的快捷菜单中选择"Generate"→"GsonFormat"插件。
③ 将服务器返回的 JSON 数据复制到 Edit 对话框中。
④ 单击"OK"按钮，就可以自动生成 Weather 类。

JSON 协议
（实践案例）

9.3.2　实践案例——访问天气实例的应用

下面我们介绍一个访问天气实例的应用。该应用从服务器获取最近一周的天气信息并显示，单击"刷新"按钮，应用重新从服务器获取数据并更新天气的数据，如图 9-32 所示。

图 9-32　天气访问实例

首先我们实现天气访问实例的界面布局，如图 9-33 所示。

```xml
<?xml version="1.0" encoding="utf-8"?>
<LinearLayout xmlns:android="http://schemas.android.com/apk/res/android"
    xmlns:app="http://schemas.android.com/apk/res-auto"
    xmlns:tools="http://schemas.android.com/tools"
    android:layout_width="match_parent"
    android:layout_height="match_parent"
    android:orientation="vertical"
    tools:context=".MainActivity">
    <TextView
        android:id="@+id/textView"
        android:layout_width="match_parent"
        android:layout_height="wrap_content"
        android:gravity="center"
        android:text="今日天气"
        android:textSize="20dp" />
    <TextView
        android:id="@+id/txtView_today"
        android:layout_width="match_parent"
        android:layout_height="wrap_content"
        android:gravity="center"
        android:text="20℃"
        android:textSize="20dp" />
    <TextView
        android:id="@+id/textView4"
        android:layout_width="match_parent"
        android:layout_height="wrap_content"
        android:gravity="center"
        android:text="本周天气"
        android:textSize="20dp" />
    <TextView
        android:id="@+id/txtView_week"
        android:layout_width="match_parent"
        android:layout_height="wrap_content"
        android:gravity="center"
        android:text="20~30℃"
        android:textSize="20dp" />
    <Button
        android:id="@+id/btn_fresh"
        android:layout_width="match_parent"
        android:layout_height="wrap_content"
        android:text="刷新" />
</LinearLayout>
```

图 9-33　天气访问实例的界面布局

天气访问实例的界面布局效果如图 9-34 所示。

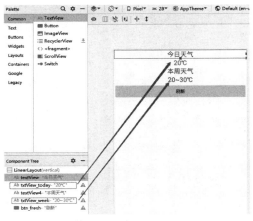

图 9-34　天气访问实例的界面布局效果

GsonFomat 插件根据 JSON 数据自动生成 Weather 类，用于数据的交互，如图 9-35 所示。

```java
package cn.edu.sziit.chapter9_jsontest.Bean;
import com.google.gson.Gson;
import com.google.gson.annotations.SerializedName;
import java.util.List;
public class Weather {
    /**
     * ERRMSG : 成功
     * WCurrent : 19
     *                                        ROWS_DETAIL :
[{"temperature":"22~27","WData":"2018-10-24"},{"temperature":"20~23","WData":"2018-10-25"},{"temperature":"20~24","WData":
"2018-10-26"},{"temperature":"20~24","WData":"2018-10-27"},{"temperature":"20~28","WData":"2018-10-28"},{"temperature":"22
~30","WData":"2018-10-29"}]
     * RESULT : S
     */
    @SerializedName("ERRMSG")
    private String ERRMSG;
    @SerializedName("WCurrent")
    private int WCurrent;
    @SerializedName("RESULT")
    private String RESULT;
    @SerializedName("ROWS_DETAIL")
    private List<ROWSDETAILBean> ROWSDETAIL;
    public static Weather objectFromData(String str) {
        return new Gson().fromJson(str, Weather.class);
    }
    public String getERRMSG() {
        return ERRMSG;
    }
    public void setERRMSG(String ERRMSG) {
        this.ERRMSG = ERRMSG;
    }
    public int getWCurrent() {
        return WCurrent;
    }
    public void setWCurrent(int WCurrent) {
        this.WCurrent = WCurrent;
    }
    public String getRESULT() {
        return RESULT;
    }
    public void setRESULT(String RESULT) {
        this.RESULT = RESULT;
    }
    public List<ROWSDETAILBean> getROWSDETAIL() {
        return ROWSDETAIL;
    }
    public void setROWSDETAIL(List<ROWSDETAILBean> ROWSDETAIL) {
        this.ROWSDETAIL = ROWSDETAIL;
    }
    public static class ROWSDETAILBean {
        /**
         * temperature : 22~27
         * WData : 2018-10-24
         */
        @SerializedName("temperature")
        private String temperature;
        @SerializedName("WData")
        private String WData;
        public static ROWSDETAILBean objectFromData(String str) {
            return new Gson().fromJson(str, ROWSDETAILBean.class);
        }
        public String getTemperature() {
```

图 9-35　Weather 类

```
            return temperature;
        }
        public void setTemperature(String temperature) {
            this.temperature = temperature;
        }
        public String getWData() {
            return WData;
        }
        public void setWData(String WData) {
            this.WData = WData;
        }
    }
}
```

图 9-35　Weather 类（续）

新建 okhttp3.Callback 回调对象并初始化，如图 9-38 所示。

```
private okhttp3.Callback mokhttp3Callback;
private void initData() {
    mokhttp3Callback = new Callback() {
        @Override
        public void onFailure(Call call, IOException e) {

        }
        @Override
        public void onResponse(Call call, final Response response)throws IOException
{
            runOnUiThread(new Runnable() {
                @Override
                public void run() {
                    //主线程进行 UI 操作
                    try {
                        parseJsonWithGson(response.body().string());
                    } catch (IOException e) {
                        e.printStackTrace();
                    }
                }
            });
        }
    };
    sendHttpRequest();
}
```

图 9-36　okhttp3.Callback 回调对象

① 新建 okhttp3.Callback 网络监听器对象。
② 初始化网络监听器对象。
③ 实现 onFailure 接口。
④ 实现 onResponse 接口。
⑤ 网络访问成功后获取数据解析。

主视图中启动 OkHttp 网络请求访问，如图 9-37 所示。
① 创建 OkHttpClienet 实例。
② 设置 POST 访问请求参数。
③ 根据网络请求的 IP 地址和 body 内容创建 Request 请求对象。
④ 使用 newCall 创建一个 Call 对象，调用 execute 发送请求获取数据，并设置回调接口为 okhttp3.Callback 监听器对象。

```java
private void sendHttpRequest() {
    String strUrl = "http://139.199.220.137:8080/api/v2/get_weather";
    /*使用OkHttp发布网络请求*/
    OkHttpClient client = new OkHttpClient();
    MediaType mediaType = MediaType.parse("application/json");
    RequestBody body = RequestBody.create(mediaType, "{\"UserName\":\"user1\"}");
    Request request = new Request.Builder()
            .url(strUrl)
            .post(body)
            .build();
    client.newCall(request).enqueue(mokhttp3Callback);
}
```

图 9-37 OkHttp 网络请求

使用 Gson 将 JSON 数据解析为 Weather 对象，提取信息并更新 UI，如图 9-38 所示。

```java
private void parseJsonWithGson(String strJsonData) {
    String strWeek="";
    Gson mGson = new Gson();
    Weather mWeather = mGson.fromJson(strJsonData, Weather.class);
    mTxtViewToday.setText("今天："+String.valueOf(mWeather.getWCurrent()) + "°C");
    for(int i=0;i<mWeather.getROWSDETAIL().size();i++)
    {
        strWeek+=mWeather.getROWSDETAIL().get(i).getWData()+":";
        strWeek+=mWeather.getROWSDETAIL().get(i).getTemperature()+"℃";
        strWeek+="\n";
    }
    mTxtViewWeek.setText(strWeek);
}
```

图 9-38 JSON 数据解析

① 新建 Gson 对象。
② 使用 Gson 将 JSON 数据转换为 Weather 对象。
③ 解析 JSON 数组，并分别解析日期和气温数据。
④ 显示一周的气温数据。

主视图的完整代码如图 9-39 所示。

```java
package cn.edu.sziit.chapter9_jsontest;
import android.os.Bundle;
import android.support.v7.app.AppCompatActivity;
import android.view.View;
import android.widget.Button;
import android.widget.TextView;
import com.google.gson.Gson;
import java.io.IOException;
import cn.edu.sziit.chapter9_jsontest.Bean.Weather;
import okhttp3.Call;
import okhttp3.Callback;
import okhttp3.MediaType;
import okhttp3.OkHttpClient;
import okhttp3.Request;
import okhttp3.RequestBody;
import okhttp3.Response;
public class MainActivity extends AppCompatActivity implements View.OnClickListener {
    private TextView mTxtViewToday;
    private TextView mTxtViewWeek;
    private okhttp3.Callback mokhttp3Callback;
    private Button mBtnFresh;
    @Override
    protected void onCreate(Bundle savedInstanceState) {
        super.onCreate(savedInstanceState);
        setContentView(R.layout.activity_main);
```

图 9-39 主视图的完整代码

```java
        initView();
        initData();
    }
    private void initData() {
        mokhttp3Callback = new Callback() {
            @Override
            public void onFailure(Call call, IOException e) {
            }

            @Override
            public void onResponse(Call call, final Response response)throws IOException {
                runOnUiThread(new Runnable() {
                    @Override
                    public void run() {
                        //主线程进行 UI 操作
                        try {
                            parseJsonWithGson(response.body().string());
                        } catch (IOException e) {
                            e.printStackTrace();
                        }
                    }
                });
            }
        };
        sendHttpRequest();
    }
    private void parseJsonWithGson(String strJsonData) {
        String strWeek="";
        Gson mGson = new Gson();
        Weather mWeather = mGson.fromJson(strJsonData, Weather.class);
        mTxtViewToday.setText("今天："+String.valueOf(mWeather.getWCurrent()) + "°C");
        for(int i=0;i<mWeather.getROWSDETAIL().size();i++)
        {
            strWeek+=mWeather.getROWSDETAIL().get(i).getWData()+":";
            strWeek+=mWeather.getROWSDETAIL().get(i).getTemperature()+"℃";
            strWeek+="\n";
        }
        mTxtViewWeek.setText(strWeek);
    }
    private void initView() {
        mTxtViewToday = (TextView) findViewById(R.id.txtView_today);
        mTxtViewWeek = (TextView) findViewById(R.id.txtView_week);
        mBtnFresh = (Button) findViewById(R.id.btn_fresh);
        mBtnFresh.setOnClickListener(this);
    }
    @Override
    public void onClick(View v) {
        switch (v.getId()) {
            case R.id.btn_fresh:
                sendHttpRequest();
                break;
        }
    }
    private void sendHttpRequest() {
        String strUrl = "http://139.199.220.137:8080/api/v2/get_weather";
        /*使用 OkHttp 发布网络请求*/
        OkHttpClient client = new OkHttpClient();
        MediaType mediaType = MediaType.parse("application/json");
        RequestBody body = RequestBody.create(mediaType, "{\"UserName\":\"user1\"}");
        Request request = new Request.Builder()
                .url(strUrl)
                .post(body)
                .build();
        client.newCall(request).enqueue(mokhttp3Callback);
    }
}
```

图 9-39 主视图的完整代码（续）

9.3.3 单元小测

判断题：

1. JSON 具备简洁和清晰的层次结构。（　　）
A. 是　　　　　　　　　　　　　　B. 否
2. JSON 数据格式可以提升网络传输效率。（　　）
A. 是　　　　　　　　　　　　　　B. 否
3. JSON 必须使用与编程语言相关的文本格式来存储和表示数据。（　　）
A. 是　　　　　　　　　　　　　　B. 否
4. 编程语言中任何支持的数据类型都可以使用 JSON 表示。（　　）
A. 是　　　　　　　　　　　　　　B. 否

选择题：

1. 关于 JSON，下面哪一项是不正确的？（　　）
A. JSON 是一种轻量级的数据交换格式
B. JSON 易于机器解析和生成
C. JSON 使用完全独立于编程语言的文本格式来存储和表示数据
D. JSON 可以提升内存的访问速度
2. OkHttp 网络访问中，网络参数代码如下：

```
RequestBody body = RequestBody.create(mediaType, "{\" UserName\ :\" user1\ "}" );
```

请问网络请求的方式是（　　）。
A. HTTP　　　　B. GET　　　　C. POST　　　　D. FTP
3. 网络回调处理的代码如下所示：

```
public void onResponse(Call call, final Response response)throws IOException
{//实现 onResponse 接口
    runOnUiThread(new Runnable() {
        public void run() {
            try {
                parseJsonWithGson(response.body().string()); //将网络成功后获取数据解析
            } catch (IOException e) {e.printStackTrace();}
        }
    });
```

请问代码中 runOnUiThread 的作用是（　　）。
A. 网络请求　　　　　　　　　　　B. 网络回调
C. 新启用一个线程　　　　　　　　D. 在 UI 主线程空闲的时候运行里面的代码
4. OkHttp 中，发送请求数据所使用的函数是（　　）。
A. execute　　　　B. loadUrl　　　　C. enqueue　　　　D. inqueue

9.4 Volley

Volley（慕课）

9.4.1 知识点讲解——Volley 网络框架

本节我们介绍 Volley 网络框架。Volley 的英文意思为"群发""迸发"。Volley 是 2013 年谷歌官方发布的一款 Android 平台上的网络通信库。Volley 非常适合一些数据量不大但需要频繁通信的网络操作。使用 Volley 进行网络开发可以使我们的开发效率得到很大的提升，而且性能的稳定性也比较高。但是 Volley 不适用于文件的上传和下载操作。

Volley 有如下特点：
- 网络通信更快，更简单，更健壮。
- GET/POST 网络请求及网络图像的高效率异步请求。
- 可以对网络请求的优先级进行排序处理。
- 可以进行网络请求的缓存。
- 可以取消多级别请求。
- 可以和 Activity 生命周期联动。

Volley 的优点主要包括：
- 非常适合数据量不大但通信频繁的网络操作。
- 直接在主线程调用服务端并处理返回结果。
- 可以取消请求，容易扩展，面向接口编程。
- 通过使用标准的 HTTP 缓存机制使磁盘和内存响应保持一致。

Volley 的缺点主要包括：
- 只支持 httpclient、HttpURLConnection。
- 对大文件下载的支持不足。
- 图片加载性能一般。

Volley 框架
（实践案例）

9.4.2 实践案例——使用 Volley 框架实现天气预报的应用

下面我们介绍一个使用 Volley 框架实现天气预报的应用。应用从服务器获取最近一周的天气信息并显示，单击"刷新"按钮，应用重新从服务器获取数据并更新，如图 9-40 所示。

图 9-40　天气访问实例

GsonFomat 可以根据 JSON 数据自动生成 Weather 类对象，如图 9-41 所示。

```java
public class Weather {
    private String ERRMSG;
    private int WCurrent;
    private String RESULT;
    private List<ROWSDETAILBean> ROWSDETAIL;
    public static Weather objectFromData(String str) {...}
    public String getERRMSG() { return ERRMSG; }
    public void setERRMSG(String ERRMSG) { this.ERRMSG = ERRMSG; }
    public int getWCurrent() { return WCurrent; }
    public void setWCurrent(int WCurrent) { this.WCurrent = WCurrent; }
    public String getRESULT() { return RESULT; }
    public void setRESULT(String RESULT) { this.RESULT = RESULT; }
    public List<ROWSDETAILBean> getROWSDETAIL() { return ROWSDETAIL; }
    public void setROWSDETAIL(List<ROWSDETAILBean> ROWSDETAIL) { this.R(
    public static class ROWSDETAILBean {
        private String temperature;
        private String WData;
        public static ROWSDETAILBean objectFromData(String str) {...}
```

图 9-41 Weather 类对象

要使用 Volley 框架和 Gson 框架，可以直接在 build.grade 的 dependencies 中增加实现 Volley 框架和 Gson 框架的路径。Volley 开源地址为'com.mcxiaoke.volley:library:1.0.19'，Gson 的开源地址为'com.google.code.gson:gson:2.8.5'，build.grade 的文件设置如图 9-42 所示。

```
apply plugin: 'com.android.application'
android {
    compileSdkVersion 27
    defaultConfig {
        applicationId "cn.edu.sziit.chapter9_volleyhttptest"
        minSdkVersion 15
        targetSdkVersion 27
        versionCode 1
        versionName "1.0"
        testInstrumentationRunner "android.support.test.runner.AndroidJUnitRunner"
    }
    buildTypes {
        release {
            minifyEnabled false
            proguardFiles getDefaultProguardFile('proguard-android.txt'), 'proguard-rules.pro'
        }
    }
}
dependencies {
    implementation fileTree(include: ['*.jar'], dir: 'libs')
    implementation 'com.android.support:appcompat-v7:27.1.1'
    implementation 'com.android.support.constraint:constraint-layout:1.1.3'
    testImplementation 'junit:junit:4.12'
    androidTestImplementation 'com.android.support.test:runner:1.0.2'
    androidTestImplementation 'com.android.support.test.espresso:espresso-core:3.0.2'
    implementation 'com.google.code.gson:gson:2.8.5'
    implementation 'com.mcxiaoke.volley:library:1.0.19'
    implementation files('libs/android-volley-master.zip')
}
```

图 9-42 build.grade 的文件设置

增加网络访问权限，配置信息文件中 Android 的网络权限为"android.permission.INTERNET"，如图 9-43 所示。

```xml
<?xml version="1.0" encoding="utf-8"?>
<manifest xmlns:android="http://schemas.android.com/apk/res/android"
    package="cn.edu.sziit.chapter9_httptest">
    <uses-permission android:name="android.permission.INTERNET" />
    <application
        android:allowBackup="true"
        android:icon="@mipmap/ic_launcher"
        android:label="@string/app_name"
        android:roundIcon="@mipmap/ic_launcher_round"
        android:supportsRtl="true"
        android:theme="@style/AppTheme">
        <activity android:name=".MainActivity">
            <intent-filter>
                <action android:name="android.intent.action.MAIN" />

                <category android:name="android.intent.category.LAUNCHER" />
            </intent-filter>
        </activity>
    </application>
</manifest>
```

图 9-43 配置 Android 的网络权限

主视图中实现主界面的监听接口，由于网络成功请求后的数据最后都是在主视图中显示的，因此监听接口需要在主视图中实现，如图 9-44 所示。

```java
//新建网络监听器对象
private com.android.volley.Response.Listener mVolleyResponseListener;
private void initData() {
//初始化网络监听器对象
    mVolleyResponseListener = new Response.Listener<JSONObject>(){
        //实现 onResponse 接口
        public void onResponse(JSONObject jsonObject) {
                parseJsonWithGson(jsonObject.toString()); //解析网络获取数据
        }
    };
}
```

图 9-44 监听接口

① 新建 volley.Response.Listener 网络监听器对象。

② 初始化网络监听器对象。

③ 实现 onResponse 接口。

④ 网络访问成功后获取数据解析。

Volley 队列初始化，如图 9-45 所示。

```java
private static RequestQueue mRequestQueue;
public static RequestQueue getRequestQueue(Context context) {
    if (mRequestQueue == null) {
        mRequestQueue = Volley.newRequestQueue(context);//如果队列为空，新建队列
    }
    return mRequestQueue;//如果全局队列不为空，直接返回 Volley 全局队列
}
```

图 9-45 Volley 队列初始化

① 新建一个 Volley 全局队列对象，如果队列为空，那么初始化全局队列。

② 如果全局队列不为空，那么直接返回 Volley 全局队列。

主视图中启动 Volley 网络请求访问，如图 9-46 所示。

① 定义访问网络的 URL 地址。

```
private void sendVolleyRequest() {
//定义访问网络的 Url 地址;
    String strUrl = "http://139.199.220.137:8080/api/v2/get_weather";
HashMap<String, String> mHashMapPara=new HashMap<>();
    mHashMapPara.put("UserName","user1"); //设置 POST 访问请求参数
    //创建 Request 请求对象; 并设置回调接口为 Volley 监听器对象
com.android.volley.Request mRequest=new JsonObjectRequest(
        com.android.volley.Request.Method.POST,
        strUrl, new Gson().toJson(mHashMapPara),
        mVolleyResponseListener,null);
    getRequestQueue(this).add(mRequest); //将请求加入 Volley 全局队列
}
```

图 9-46　Volley 队列初始化

② 设置 POST 访问请求参数。

③ 根据网络请求的 IP 地址和 body 内容创建 Request 请求对象,并设置回调接口为 Volley 监听器对象。

④ 将请求加入 Volley 全局队列。

使用 Gson 将 JSON 数据解析为 Weather 对象,提取天气信息并更新主视图的 UI 组件,如图 9-47 所示。

```
private void parseJsonWithGson(String strJsonData) {
    String strWeek="";
    Gson mGson = new Gson(); //新建 Gson 对象
//使用 Gson 将 Json 数据转换为 Weather 对象
    Weather mWeather = mGson.fromJson(strJsonData, Weather.class);
mTxtViewToday.setText("今天:"+mWeather.getWCurrent() + "℃");
    //解析 Json 数组;并分别解析日期和气温数据
    for(int i=0;i<mWeather.getROWSDETAIL().size();i++)
    {
        strWeek+=mWeather.getROWSDETAIL().get(i).getWData()+":"; //解析日期数据
//解析气温数据
        strWeek+=mWeather.getROWSDETAIL().get(i).getTemperature()+"℃";
strWeek+= "\n";
    }
    mTxtViewWeek.setText(strWeek);//显示一周的气温数据
}
```

图 9-47　Gson 数据解析

① 新建 Gson 对象。

② 使用 Gson 将 JSON 数据转换为 Weather 对象。

③ 解析 JSON 数组,并分别解析日期和气温数据。

④ 显示一周的气温数据。

主视图的完整代码如图 9-48 所示。

```
package cn.edu.sziit.chapter9_volleyhttptest;
import android.content.Context;
import android.os.Bundle;
import android.support.v7.app.AppCompatActivity;
import android.view.View;
import android.widget.Button;
import android.widget.TextView;
import com.android.volley.RequestQueue;
import com.android.volley.Response;
import com.android.volley.toolbox.JsonObjectRequest;
import com.android.volley.toolbox.Volley;
import com.google.gson.Gson;
import org.json.JSONObject;
```

图 9-48　主视图的完整代码

```java
import java.util.HashMap;
import cn.edu.sziit.chapter9_volleyhttptest.Bean.Weather;
public class MainActivity extends AppCompatActivity implements View.OnClickListener {
    private TextView mTxtViewToday;
    private TextView mTxtViewWeek;
    private com.android.volley.Response.Listener mVolleyResponseListener;
    private Button mBtnFresh;
    private static RequestQueue mRequestQueue;
    @Override
    protected void onCreate(Bundle savedInstanceState) {
        super.onCreate(savedInstanceState);
        setContentView(R.layout.activity_main);
        initView();
        initData();
    }
    private void initData() {
        mVolleyResponseListener = new Response.Listener<JSONObject>(){
            @Override
            public void onResponse(JSONObject jsonObject) {
                parseJsonWithGson(jsonObject.toString());
            }
        };
        sendVolleyRequest();
    }
    private void parseJsonWithGson(String strJsonData) {
        String strWeek="";
        Gson mGson = new Gson();
        Weather mWeather = mGson.fromJson(strJsonData, Weather.class);
        mTxtViewToday.setText("今天："+String.valueOf(mWeather.getWCurrent()) + "° C");
        for(int i=0;i<mWeather.getROWSDETAIL().size();i++)
        {
            strWeek+=mWeather.getROWSDETAIL().get(i).getWData()+":";
            strWeek+=mWeather.getROWSDETAIL().get(i).getTemperature()+"℃";
            strWeek+="\n";
        }
        mTxtViewWeek.setText(strWeek);
    }
    private void initView() {
        mTxtViewToday = (TextView) findViewById(R.id.txtView_today);
        mTxtViewWeek = (TextView) findViewById(R.id.txtView_week);
        mBtnFresh = (Button) findViewById(R.id.btn_fresh);
        mBtnFresh.setOnClickListener(this);
    }
    @Override
    public void onClick(View v) {
        switch (v.getId()) {
            case R.id.btn_fresh:
                sendVolleyRequest();
                break;
        }
    }
    public static RequestQueue getRequestQueue(Context context) {
        if (mRequestQueue == null) {
            mRequestQueue = Volley.newRequestQueue(context);
        }
        return mRequestQueue;
    }
    private void sendVolleyRequest() {
        String strUrl = "http://139.199.220.137:8080/api/v2/get_weather";
        HashMap<String, String> mHashMapPara=new HashMap<>();
        mHashMapPara.put("UserName","user1");
        /*使用 Volley 发布网络请求*/
        com.android.volley.Request mRequest=new JsonObjectRequest(
                com.android.volley.Request.Method.POST,
                strUrl, new Gson().toJson(mHashMapPara),
                mVolleyResponseListener,null);
        getRequestQueue(this).add(mRequest);
    }
}
```

图 9-48 主视图的完整代码（续）

9.4.3 单元小测

判断题：

1. Volley 适合数据量大并且通信频繁的操作。（　　）
   A. 是　　　　　　　　　　　　　　　B. 否
2. Volley 适合下载大容量的文件。（　　）
   A. 是　　　　　　　　　　　　　　　B. 否
3. Volley 对于图片加载有很好的性能。（　　）
   A. 是　　　　　　　　　　　　　　　B. 否
4. Volley 使用标准的 HTTP 缓存机制保持磁盘和内存响应的一致。（　　）
   A. 是　　　　　　　　　　　　　　　B. 否
5. Volley 访问网络不需要创建 Volley 对象。（　　）
   A. 是　　　　　　　　　　　　　　　B. 否

选择题：

1. 关于 Volley，下面哪一项是不正确的？（　　）
   A. Volley 非常适合数据量不大但通信频繁的网络操作
   B. Volley 直接在主线程调用服务端并处理返回结果
   C. Volley 可以取消请求，容易扩展，面向接口编程
   D. Volley 是第三方框架，不支持 HttpURLConnection
2. com.android.volley.Response.Listener 监听接口必须实现的方法是（　　）。
   A. onFail()　　　　　　　　　　　　B. onFailure()
   C. onResponse()　　　　　　　　　　D. onSuccess()
3. Volley 发送网络请求的代码如下所示：

```
private void sendVolleyRequest() {
    String strUrl = "http://139.199.220.137:8080/api/v2/get_weather"; //定义访问网络的 Url 地址;
    HashMap<String, String> mHashMapPara=new HashMap<>();
    mHashMapPara.put("UserName","user1");
    com.android.volley.Request mRequest=new JsonObjectRequest(
        com.android.volley.Request.Method.POST,
        strUrl,
        new Gson().toJson(mHashMapPara),
        mVolleyResponseListener,null);
}
```

请问 Volley 网络请求的类型是（　　）。
A. HTTP　　　　　　　　　　　　　B. GET
C. POST　　　　　　　　　　　　　D. FTP

4. Volley 发送网络请求的代码如下所示，请补全代码。（　　）

```
private void sendVolleyRequest() {
    String strUrl = "http://139.199.220.137:8080/api/v2/get_weather"; //定义访问网络的 Url 地址;
    HashMap<String, String> mHashMapPara=new HashMap<>();
    mHashMapPara.put("UserName","user1");
    com.android.volley.Request mRequest=new JsonObjectRequest(
```

```
            com.android.volley.Request.Method.POST,
            strUrl,
            new Gson().toJson(mHashMapPara),
            mVolleyResponseListener,null);
            (    ?    )
}
```

A. getRequestQueue(this).execute(mRequest)

B. getRequestQueue(this).add(mRequest)

C. getRequestQueue(this).enqueue(mRequest)

D. getRequestQueue(this).execute(mRequest)

本章课后练习和程序源代码

第 9 章源代码及课后习题

# 10 数据存储

 知识点

Android 的文件存储、Android 的数据库存储。

 能力点

1. 熟练应用 Android 的文件存储服务。
2. 熟练使用 Android 的 SharePreferences 方式存储信息。
3. 熟练使用 SQLite 轻量级框架进行数据库存储。

文件存储（慕课）

 ## 10.1 文件存储

本节我们介绍文件存储，如果 App 内有些数据需要使用到上次该 App 关闭时的数据，又比如下次启动 App 没有网络时要求显示之前的省市信息，那么进行本地存储就显得非常有必要。本地存储就是将内存瞬时数据保存到存储设备变为持久化的数据，Android 数据存储主要有 4 种方式，如图 10-1 所示。

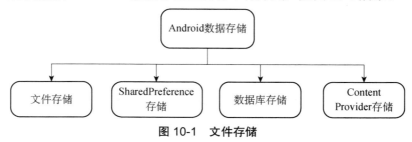

图 10-1 文件存储

- ◆ 文件存储：数据通过 I/O 存储为文件，用于存储大数量的文本或者二进制数据。
- ◆ SharedPreference 存储：数据通过 XML 文件存储，用于存储简单的参数设置。
- ◆ 数据库存储：数据通过轻量级的 SQLite 文件存储，用于存储复杂而关联的数据。
- ◆ ContentProvider 存储：应用程序数据共享存储，用于存储音/视频和通讯录等系统数据。

文件存储是 Android 中最基本的一种数据存储方式，它不对存储的内容进行任何的格式化处理，所有数据都原封不动地保存到文件当中。它比较适合存储一些简单的文本数据或二进制数据。如果你想使用文件存储方式来保存一些较为复杂的文本数据，就需要定义一套自己的格式规范，方便之后将文件重新解析出来。

Java 提供 I/O 流存储方式，FileOutputStream 提供 openFileOutPut 方法将数据写入文件，FileInputStream 提供 openFileInPut 方法读取文件，Android 提供 Enviroment 类对 Android 设备的 SD 卡进行文件数据的读写。Enviroment 类的方法如表 10-1 所示。

表 10-1　Enviroment 类的方法

序号	类方法	方法说明
1	getRootDirectory	获取系统根目录路径
2	getDataDirectory	获取系统数据目录路径
3	getDownloadCacheDirectory	获取下载缓存目录路径
4	getExternalStorageDirectory	获取外部存储（SD 卡）目录路径
5	getExternalStorageState	获取外部存储（SD 卡）的状态
6	getStorageState	获取指定目录的状态

图 10-2 所示的是一个文件存储应用读取用户注册信息的实例。文件存储应用中输入学生的姓名、学号、班级和爱好后，单击"保存文本到 SD 卡"按钮，显示用户文件的保存路径；文件读取应用中将 SD 下所有的 TXT 文本列表显示，选择一个学生文件后，显示用户的学生的姓名、学号、班级和爱好信息。

图 10-2　文件存储

10.1.1　文件保存

本小节我们介绍文件的保存功能，输入学生的姓名、学号、班级和爱好后，单击"保存文本到 SD 卡"按钮，将学生的信息保存到文件中，并显示用户文件的保存路径。

文件读写布局
（实践案例）

首先我们完成应用的界面布局 activity_text_write.xml，如图 10-3 所示。

```xml
<LinearLayout xmlns:android="http://schemas.android.com/apk/res/android"
    android:layout_width="match_parent"
    android:layout_height="match_parent"
    android:focusable="true"
    android:focusableInTouchMode="true"
    android:orientation="vertical"
    android:padding="10dp" >
    <RelativeLayout
        android:layout_width="match_parent"
        android:layout_height="50dp" >
        <TextView
            android:id="@+id/tv_name"
            android:layout_width="wrap_content"
            android:layout_height="match_parent"
            android:layout_alignParentLeft="true"
            android:gravity="center"
            android:text="姓名："
            android:textColor="@color/black"
            android:textSize="17sp" />
        <EditText
            android:id="@+id/et_name"
            android:layout_width="match_parent"
            android:layout_height="match_parent"
            android:layout_marginBottom="5dp"
            android:layout_marginTop="5dp"
            android:layout_toRightOf="@+id/tv_name"
            android:background="@drawable/edittext_selector"
            android:gravity="left|center"
            android:hint="请输入姓名"
            android:inputType="text"
            android:maxLength="12"
            android:textColor="@color/black"
            android:textColorHint="@color/grey"
            android:textCursorDrawable="@drawable/text_cursor"
            android:textSize="17sp" />
    </RelativeLayout>
    <RelativeLayout
        android:layout_width="match_parent"
        android:layout_height="50dp" >
        <TextView
            android:id="@+id/tv_no"
            android:layout_width="wrap_content"
            android:layout_height="match_parent"
            android:layout_alignParentLeft="true"
            android:gravity="center"
            android:text="学号："
            android:textColor="@color/black"
            android:textSize="17sp" />
        <EditText
            android:id="@+id/et_no"
            android:layout_width="match_parent"
            android:layout_height="match_parent"
            android:layout_marginTop="5dp"
            android:layout_marginBottom="5dp"
            android:layout_toRightOf="@+id/tv_no"
            android:background="@drawable/edittext_selector"
            android:gravity="left|center"
            android:hint="请输入学号"
            android:inputType="text"
            android:maxLength="20"
            android:textColor="@color/black"
            android:textColorHint="@color/grey"
            android:textCursorDrawable="@drawable/text_cursor"
            android:textSize="17sp" />
    </RelativeLayout>
    <RelativeLayout
```

图 10-3　文件保存界面布局

```xml
        android:layout_width="match_parent"
        android:layout_height="50dp" >
        <TextView
            android:id="@+id/tv_class"
            android:layout_width="wrap_content"
            android:layout_height="match_parent"
            android:layout_alignParentLeft="true"
            android:gravity="center"
            android:text="班级："
            android:textColor="@color/black"
            android:textSize="17sp" />
        <EditText
            android:id="@+id/et_class"
            android:layout_width="match_parent"
            android:layout_height="match_parent"
            android:layout_marginTop="5dp"
            android:layout_marginBottom="5dp"
            android:layout_toRightOf="@+id/tv_class"
            android:background="@drawable/editext_selector"
            android:gravity="left|center"
            android:hint="请输入班级"
            android:inputType="text"
            android:maxLength="20"
            android:textColor="@color/black"
            android:textColorHint="@color/grey"
            android:textCursorDrawable="@drawable/text_cursor"
            android:textSize="17sp" />
    </RelativeLayout>
    <RelativeLayout
        android:layout_width="match_parent"
        android:layout_height="50dp" >
        <TextView
            android:id="@+id/tv_hobby"
            android:layout_width="wrap_content"
            android:layout_height="match_parent"
            android:layout_alignParentLeft="true"
            android:gravity="center"
            android:text="爱好："
            android:textColor="@color/black"
            android:textSize="17sp" />
        <EditText
            android:id="@+id/et_hobby"
            android:layout_width="match_parent"
            android:layout_height="match_parent"
            android:layout_marginTop="5dp"
            android:layout_marginBottom="5dp"
            android:layout_toRightOf="@+id/tv_hobby"
            android:background="@drawable/editext_selector"
            android:gravity="left|center"
            android:hint="请输入爱好:"
            android:inputType="text"
            android:maxLength="20"
            android:textColor="@color/black"
            android:textColorHint="@color/grey"
            android:textCursorDrawable="@drawable/text_cursor"
            android:textSize="17sp" />
    </RelativeLayout>
    <Button
        android:id="@+id/btn_save"
        android:layout_width="match_parent"
        android:layout_height="wrap_content"
        android:text="保存文本到 SD 卡"
        android:textColor="@color/black"
        android:textSize="20sp" />
    <TextView
```

图 10-3　文件保存界面布局（续）

```
            android:id="@+id/tv_path"
            android:layout_width="wrap_content"
            android:layout_height="match_parent"
            android:textColor="@color/black"
            android:textSize="17sp" />
</LinearLayout>
```

图 10-3 文件保存界面布局（续）

文件保存界面布局的效果如图 10-4 所示。

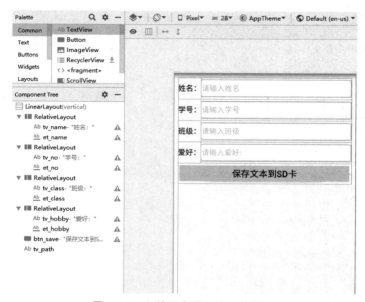

图 10-4 文件保存界面布局的效果

下面实现申请访问 SD 外部存储权限。

① 首先检查应用是否有读取 SD 卡权限，如果没有的话需要向用户申请权限，如图 10-5 所示。

```
private void getSdPermission() {
//首先检查应用是否有读取 SD 卡权限
    if (ContextCompat.checkSelfPermission(this,
Manifest.permission.READ_EXTERNAL_STORAGE) !=          PackageManager.PERMISSION_GRANTED) {
//申请权限
ActivityCompat.requestPermissions(this, new
String[]{Manifest.permission.READ_EXTERNAL_STORAGE}, 1);
}
//检查应用是否有写 SD 卡权限
    if (ContextCompat.checkSelfPermission(this,
Manifest.permission.WRITE_EXTERNAL_STORAGE)!=
PackageManager.PERMISSION_GRANTED) {
   //申请权限
            ActivityCompat.requestPermissions(this, new
String[]{Manifest.permission.WRITE_EXTERNAL_STORAGE}, 1);
}
}
```

图 10-5 申请读取 SD 卡权限

② 权限申请的处理，如图 10-6 所示。Activity 中重写 onRequestPermissionsResult() 方法

```java
public void onRequestPermissionsResult(int requestCode, @NonNull String[] permissions, @NonNull int[] grantResults) {
    switch (requestCode) {
        case 1:
            if (grantResults.length > 0 && grantResults[0] == PackageManager.PERMISSION_GRANTED) {
            } else {
                Toast.makeText(this, "you denied the permission", Toast.LENGTH_SHORT).show();}
            break; }
}
```

图 10-6 onRequestPermissionsResult 处理回调

处理权限结果回调。当用户处理完授权操作时，系统会自动回调该方法，该方法有三个参数。

- int requestCode：在调用 requestPermissions() 时的第一个参数。
- String[] permissions：权限数组，在调用 requestPermissions() 时的第二个参数。
- int[] grantResults：授权结果数组。

③ 增加网络访问权限，配置信息文件中 Android 的网络权限为 "android.permission.INTERNET" 和 ="android.permission.READ_EXTERNAL_STORAGE"，如图 10-7 所示。

```xml
<?xml version="1.0" encoding="utf-8"?>
<manifest xmlns:android="http://schemas.android.com/apk/res/android"
    package="com.example.chapter10file">

    <uses-permission android:name="android.permission.WRITE_EXTERNAL_STORAGE" />
    <uses-permission android:name="android.permission.READ_EXTERNAL_STORAGE" />
    <application
        android:allowBackup="true"
        android:icon="@mipmap/ic_launcher"
        android:label="@string/app_name"
        android:roundIcon="@mipmap/ic_launcher_round"
        android:supportsRtl="true"
        android:theme="@style/AppTheme">
        <activity android:name=".MainActivity">
            <intent-filter>
                <action android:name="android.intent.action.MAIN" />

                <category android:name="android.intent.category.LAUNCHER" />
            </intent-filter>
        </activity>
    </application>
</manifest>
```

图 10-7 配置 Android 的网络权限

Android 中将字符串内容保存到文件的代码，如图 10-8 所示。使用路径参数新建文件输出流，将文本通过输出流保存到文件，关闭输出流。

```java
public static void saveText(String path, String txt) {
    try {
        //使用路径参数新建文件输出流；
        FileOutputStream fos = new FileOutputStream(path);        fos.write(txt.getBytes()); //将文本通过输出流保存到文件
        fos.close(); //关闭输出流
    } catch (Exception e) {
        e.printStackTrace();
    }
}
```

图 10-8 字符串保存到文件

将数据保存为文件的时候，需要对文件进行命名，一般使用保存文件的时间来对文件进行唯一识别，获取系统时间的代码，如图 10-9 所示，新建时间格式化对象，格式化当前时间。

```java
public static String getNowDateTime(String formatStr) {
    String format = formatStr;
    if (format==null || format.length()<=0) {
        format = "yyyyMMddHHmmss";
    }
    SimpleDateFormat s_format = new SimpleDateFormat(format);//新建时间格式化对象
    return s_format.format(new Date()); //使用对象格式化当前时间
}
```

图 10-9　获取当前时间

将学生的信息保存到文件的代码，如图 10-10 所示。首先判断 SD 卡是否可以读取，如果可以读取初始化文件路径，则将数据保存到文件，在视图组件中显示文件信息。

```java
public static String saveDataToFile() {
//首先判断 SD 卡是否可以读取
    if (Environment.getExternalStorageState().equals(Environment.MEDIA_MOUNTED)
 == true) {
        //初始化文件路径
        String file_path = Environment.getExternalStorageDirectory() + "/"
+getNowDateTime("") +".txt";
saveText(file_path, content); //将数据保存到文件
//在视图组件显示文件信息
        mTvPath.setText("用户注册信息文件的保存路径为：\n"+file_path);
Toast.makeText(this, "数据已写入 SD 卡文件：", Toast.LENGTH_SHORT).show();
} else {
        Toast.makeText(this, "未发现 SD 卡，请检查：", Toast.LENGTH_SHORT).show();
}
}
```

图 10-10　学生的信息保存到文件

新建文件保存视图 TextWriteActivity，加载布局文件 activity_text_write.XML。文件保存视图 TextWriteActivity 的整体代码如图 10-11 所示。

```java
package com.example.chapter10file;
import android.Manifest;
import android.content.pm.PackageManager;
import android.os.Bundle;
import android.os.Environment;
import android.support.annotation.NonNull;
import android.support.v4.app.ActivityCompat;
import android.support.v4.content.ContextCompat;
import android.support.v7.app.AppCompatActivity;
import android.text.TextUtils;
import android.view.View;
import android.view.View.OnClickListener;
import android.widget.Button;
import android.widget.EditText;
import android.widget.TextView;
import android.widget.Toast;
import com.example.chapter10file.util.DateUtil;
import com.example.chapter10file.util.FileUtil;
public class TextWriteActivity extends AppCompatActivity implements OnClickListener {
    private EditText mEtName;
    private EditText mEtNo;
    private EditText mEtClass;
    private EditText mEtHobby;
    private Button mBtnSave;
```

图 10-11　TextWriteActivity 代码

```java
    private TextView mTvPath;
    private String strFilePath;
    @Override
    protected void onCreate(Bundle savedInstanceState) {
        super.onCreate(savedInstanceState);
        setContentView(R.layout.activity_text_write);
        initView();
        initData();
    }
    private void initData() {
        strFilePath=Environment.getExternalStorageDirectory() + "/";
    }
    @Override
    public void onClick(View v) {
        switch (v.getId()) {
            case R.id.btn_save:
                getSdPermission();
                submit();
                break;
            default:
                break;
        }
    }
    private void initView() {
        mEtName = (EditText) findViewById(R.id.et_name);
        mEtNo = (EditText) findViewById(R.id.et_no);
        mEtClass = (EditText) findViewById(R.id.et_class);
        mEtHobby = (EditText) findViewById(R.id.et_hobby);
        mBtnSave = (Button) findViewById(R.id.btn_save);
        mTvPath = (TextView) findViewById(R.id.tv_path);
        mBtnSave.setOnClickListener(this);
    }
    private void submit() {
        // validate
        String name = mEtName.getText().toString().trim();
        if (TextUtils.isEmpty(name)) {
            Toast.makeText(this, "请输入姓名", Toast.LENGTH_SHORT).show();
            return;
        }
        String no = mEtNo.getText().toString().trim();
        if (TextUtils.isEmpty(no)) {
            Toast.makeText(this, "请输入学号", Toast.LENGTH_SHORT).show();
            return;
        }
        String cs =mEtClass.getText().toString().trim();
        if (TextUtils.isEmpty( cs)){
            Toast.makeText(this, "请输入班级", Toast.LENGTH_SHORT).show();
            return;
        }
        String hobby = mEtHobby.getText().toString().trim();
        if (TextUtils.isEmpty(hobby)) {
            Toast.makeText(this, "请输入爱好:", Toast.LENGTH_SHORT).show();
            return;
        }
        String content = "";
        content = String.format("%s 姓名: %s\n", content, name);
        content = String.format("%s 学号: %s\n", content, no);
        content = String.format("%s 班级: %s\n", content, cs);
        content = String.format("%s 爱好: %s\n", content, hobby);
        content = String.format("%s 注册时间: %s\n", content, DateUtil.getNowDateTime("yyyy-MM-dd HH:mm:ss"));
        if (Environment.getExternalStorageState().equals(Environment.MEDIA_MOUNTED) == true) {
            String file_path = strFilePath + DateUtil.getNowDateTime("") + ".txt";
            FileUtil.saveText(file_path, content);
            mTvPath.setText("用户注册信息文件的保存路径为: \n"+file_path);
            Toast.makeText(this, "数据已写入 SD 卡文件:", Toast.LENGTH_SHORT).show();
```

图 10-11 TextWriteActivity 代码（续）

```java
        } else {
            Toast.makeText(this, "未发现 SD 卡，请检查:", Toast.LENGTH_SHORT).show();
        }
    }
    private void getSdPermission() {
        if (ContextCompat.checkSelfPermission(this, Manifest.permission.READ_EXTERNAL_STORAGE) != PackageManager.PERMISSION_GRANTED) {
            ActivityCompat.requestPermissions(this, new String[]{Manifest.permission.READ_EXTERNAL_STORAGE}, 1);
        }
        if (ContextCompat.checkSelfPermission(this, Manifest.permission.WRITE_EXTERNAL_STORAGE) != PackageManager.PERMISSION_GRANTED) {
            ActivityCompat.requestPermissions(this, new String[]{Manifest.permission.WRITE_EXTERNAL_STORAGE}, 1);
        }
    }
    @Override
    public void onRequestPermissionsResult(int requestCode, @NonNull String[] permissions, @NonNull int[] grantResults) {
        switch (requestCode) {
            case 1:
                if (grantResults.length > 0 && grantResults[0] == PackageManager.PERMISSION_GRANTED) {
                } else {
                    Toast.makeText(this, "you denied the permission", Toast.LENGTH_SHORT).show();
                }
                break;
            default:
        }
    }
}
```

图 10-11　TextWriteActivity 代码（续）

10.1.2　文件读取

文件读取（实践案例）

本小节我们介绍文件的读取功能，将 SD 下所有的 TXT 文本列表显示，选择一个学生文件后，显示学生的姓名、学号、班级和爱好信息。首先我们完成应用的界面布局 activity_text_read.xml，如图 10-12 所示。

```xml
<LinearLayout xmlns:android="http://schemas.android.com/apk/res/android"
    android:layout_width="match_parent"
    android:layout_height="match_parent"
    android:focusable="true"
    android:focusableInTouchMode="true"
    android:orientation="vertical"
    android:padding="10dp" >
<Button
    android:id="@+id/btn_delete"
    android:layout_width="match_parent"
    android:layout_height="wrap_content"
    android:text="删除所有文本文件"
    android:textColor="@color/black"
    android:textSize="20sp" />
<RelativeLayout
    android:layout_width="match_parent"
    android:layout_height="50dp" >
    <TextView
        android:id="@+id/tv_file"
        android:layout_width="wrap_content"
        android:layout_height="match_parent"
        android:layout_alignParentLeft="true"
        android:gravity="center"
        android:text="文件名："
        android:textColor="@color/black"
```

图 10-12　文件读写界面布局

```xml
                    android:textSize="17sp" />
            <Spinner
                android:id="@+id/sp_file"
                android:layout_width="match_parent"
                android:layout_height="match_parent"
                android:layout_toRightOf="@+id/tv_file"
                android:gravity="left|center"
                ></Spinner>
    </RelativeLayout>
    <TextView
        android:id="@+id/tv_filecontent"
        android:layout_width="match_parent"
        android:layout_height="wrap_content"
        android:textColor="@color/black"
        android:textSize="17sp" />
</LinearLayout>
```

图 10-12　文件读写界面布局（续）

文件读写界面布局的效果如图 10-13 所示。

图 10-13　文件读写界面效果

从文件中读取内容并保存到字符串的实现代码，如图 10-14 所示，文件读取的流程如下：

```java
public static String openText(String path) {
    String readStr = "";
    try {FileInputStream fis = new FileInputStream(path); //新建文件输出流
        byte[] b = new byte[fis.available()];//新建字节数组
        fis.read(b); //将文件内容通过输入流拷贝到字节数组
        readStr = new String(b);//将字节数组内容转换为字符串
        fis.close(); //关闭输入流
    } catch (Exception e) {
e.printStackTrace();
}
    return readStr;
}
```

图 10-14　文件读取功能

① 新建文件输出流。
② 新建字节数组。
③ 将文件内容通过输入流复制到字节数组中。
④ 将字节数组内容转换为字符串。
⑤ 关闭输入流。

文件系统的文件非常多，我们需要获取的是后缀名为 txt 的文件，获取文件夹的所有指定后缀名的文件的代码，如图 10-15 所示。

获取文件夹的指定文件的流程如下：

```java
public static ArrayList<File> getFileList(String path, String[] extendArray) {
    ArrayList<File> displayedContent = new ArrayList<File>();//新建文件列表对象
    File[] files = null; //新建文件对象数组
    File directory = new File(path); //新建文件路径对象
    if (extendArray != null && extendArray.length>0) {
        FilenameFilter fileFilter = getTypeFilter(extendArray); //新建文件过滤器对象
        files = directory.listFiles(fileFilter); //文件夹中过滤后缀名后存入文件对象数组
    } else {
        files = directory.listFiles();
    }
    if (files != null) {
//将文件对象数组中的文件存储到文件列表对象
        for (File f : files) {
            if (!f.isDirectory() && !f.isHidden()) {
                displayedContent.add(f);
            }
        }
    }
    return displayedContent;
}
```

图 10-15　获取文件夹的所有指定后缀名文件

① 新建文件列表对象。
② 新建文件对象数组。
③ 新建文件路径对象。
④ 新建文件过滤器对象。
⑤ 将文件过滤后缀名后的结果存入文件对象数组。
⑥ 将文件对象数组中的文件存储到文件列表对象。

获取后缀名为 txt 的文件后，将指定文件夹的所有 txt 文档显示到 Spinner 控件，实现代码如图 10-16 所示。获取文件列表，为文件名数组分配内存，给文件名数组赋值，新建适配器，为 Spinner 控件设置适配器。

```java
private void refreshSpinner() {
//获取文件列表
    ArrayList<File> mFileLists =
getFileList(Environment.getExternalStorageDirectory() + "/";, new String[]{".txt"});
    if (mFileLists.size() > 0) {
        strFilethisNames = new String[mFileLists.size()]; //文件名数组分配内存
        for (int i = 0; i < mFileLists.size(); i++) {
//给文件名数组赋值
            strFilethisNames[i] = mFileLists.get(i).getName();
        }
//新建适配器
        ArrayAdapter<String> adapter = new ArrayAdapter<String>(this,
                android.R.layout.simple_spinner_item, strFilethisNames);
adapter.setDropDownViewResource(android.R.layout.simple_spinner_dropdown_item);
        mSpFile.setAdapter(adapter); //Spinner 控件设置适配器
    mSpFile.setSelection(0);
    }
}
```

图 10-16　Spinner 控件数据显示

单击 Spinner 控件，选择指定的文本记录的文件，事件处理的代码如图 10-17 所示，Spinner

控件的处理流程如下：

```
public void onItemSelected(AdapterView<?> parent, View view, int position, long id) {
    String file_path = strFilePath + ((Spinner) parent).getSelectedItem().toString();//获取文件路径
    String content = FileUtil.openText(file_path);//读取文件内容到字符串
    //组件中显示字符串内容
    mTvFilecontent.setText("读取文件内容如下：\n" + content);
}
```

图 10-17　Spinner 响应事件

① 获取文件路径。
② 读取文件内容到字符串。
③ 在组件中显示字符串内容。

新建文件读取视图 TextReadActivity，加载布局文件 activity_text_read.xml，文件读取视图 TextReadActivity 的整体代码如图 10-18 所示。

```java
package com.example.chapter10file;
import android.Manifest;
import android.content.pm.PackageManager;
import android.os.Bundle;
import android.os.Environment;
import android.support.annotation.NonNull;
import android.support.v4.app.ActivityCompat;
import android.support.v4.content.ContextCompat;
import android.support.v7.app.AppCompatActivity;
import android.view.View;
import android.view.View.OnClickListener;
import android.widget.AdapterView;
import android.widget.ArrayAdapter;
import android.widget.Button;
import android.widget.Spinner;
import android.widget.TextView;
import android.widget.Toast;
import com.example.chapter10file.util.FileUtil;
import java.io.File;
import java.util.ArrayList;
public class TextReadActivity extends AppCompatActivity implements OnClickListener, AdapterView.OnItemSelectedListener {
    private Button mBtnDelete;
    private Spinner mSpFile;
    private TextView mTvFilecontent;
    private String strFilePath;
    String[] strFilethisNames;
    @Override
    protected void onCreate(Bundle savedInstanceState) {
        super.onCreate(savedInstanceState);
        setContentView(R.layout.activity_text_read);
        initView();
        initData();
    }
    private void initData() {
        strFilePath = Environment.getExternalStorageDirectory() + "/";
        if (Environment.getExternalStorageState().equals(Environment.MEDIA_MOUNTED) == true) {
            refreshSpinner();
        } else {
            Toast.makeText(this, "未发现已挂载的 SD 卡，请检查", Toast.LENGTH_SHORT).show();
        }
    }
    private void refreshSpinner() {
        getSdPermission();
        ArrayList<File> mFileLists = FileUtil.getFileList(strFilePath, new String[]{".txt"});
        if (mFileLists.size() > 0) {
            strFilethisNames = new String[mFileLists.size()];
            for (int i = 0; i < mFileLists.size(); i++) {
```

图 10-18　TextReadActivity 代码

```java
                    strFilethisNames[i] = mFileLists.get(i).getName();
                }
                ArrayAdapter<String> adapter = new ArrayAdapter<String>(this,
                        android.R.layout.simple_spinner_item,                                  strFilethisNames);
adapter.setDropDownViewResource(android.R.layout.simple_spinner_dropdown_item);
                mSpFile.setPrompt("请选择文本文件");
                mSpFile.setAdapter(adapter);
                mSpFile.setSelection(0);
            }
        }
        @Override
        public void onClick(View v) {
            if (v.getId() == R.id.btn_delete) {
                for (int i = 0; i < strFilethisNames.length; i++) {
                    String file_path = strFilePath + strFilethisNames[i];
                    File f = new File(file_path);
                    boolean result = f.delete();
                    if (result != true) {
                        Toast.makeText(this, "文件删除不成功", Toast.LENGTH_SHORT).show();
                    }
                }
                refreshSpinner();
                Toast.makeText(this, "已删除临时目录下的所有文本文件", Toast.LENGTH_SHORT).show();
            }
        }
        private void initView() {
            mBtnDelete = (Button) findViewById(R.id.btn_delete);
            mSpFile = (Spinner) findViewById(R.id.sp_file);
            mTvFilecontent = (TextView) findViewById(R.id.tv_filecontent);
            mBtnDelete.setOnClickListener(this);
            mSpFile.setOnItemSelectedListener(this);
        }
        @Override
        public void onItemSelected(AdapterView<?> parent, View view, int position, long id) {
            String file_path = strFilePath + ((Spinner) parent).getSelectedItem().toString();
            String content = FileUtil.openText(file_path);
            mTvFilecontent.setText("读取文件内容如下：\n" + content);
        }
        @Override
        public void onNothingSelected(AdapterView<?> parent) {

        }
        private void getSdPermission() {
            if     (ContextCompat.checkSelfPermission(this,         Manifest.permission.READ_EXTERNAL_STORAGE)        !=
PackageManager.PERMISSION_GRANTED) {
                ActivityCompat.requestPermissions(this, new String[]{Manifest.permission.READ_EXTERNAL_STORAGE}, 1);
            }
            if     (ContextCompat.checkSelfPermission(this,         Manifest.permission.WRITE_EXTERNAL_STORAGE)        !=
PackageManager.PERMISSION_GRANTED) {
                ActivityCompat.requestPermissions(this, new String[]{Manifest.permission.WRITE_EXTERNAL_STORAGE}, 1);
            }
        }
        @Override
        public void onRequestPermissionsResult(int requestCode, @NonNull String[] permissions, @NonNull int[] grantResults) {
            switch (requestCode) {
                case 1:
                    if (grantResults.length > 0 && grantResults[0] == PackageManager.PERMISSION_GRANTED) {

                    } else {
                        Toast.makeText(this, "you denied the permission", Toast.LENGTH_SHORT).show();
                    }
                    break;

                default:
            }
        }
    }
```

图 10-18 TextReadActivity 代码（续）

使用文件管理器查看保存文件地址，如图 10-19 所示，进入 SD 卡目录，可以看到已保存信息的文件，可以下载并查看文件的内容。

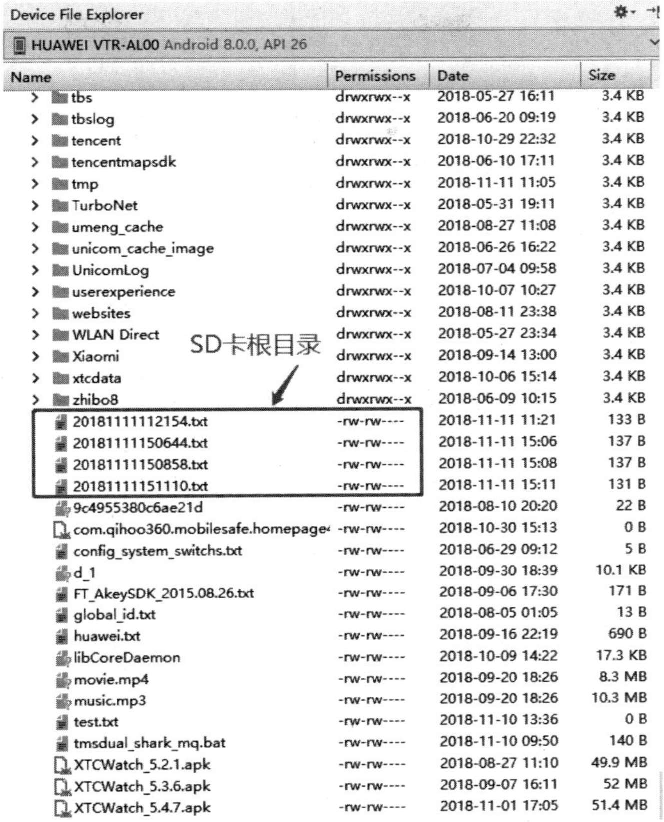

Name	Permissions	Date	Size
tbs	drwxrwx--x	2018-05-27 16:11	3.4 KB
tbslog	drwxrwx--x	2018-06-20 09:19	3.4 KB
tencent	drwxrwx--x	2018-10-29 22:32	3.4 KB
tencentmapsdk	drwxrwx--x	2018-06-10 17:11	3.4 KB
tmp	drwxrwx--x	2018-11-11 11:05	3.4 KB
TurboNet	drwxrwx--x	2018-05-31 19:11	3.4 KB
umeng_cache	drwxrwx--x	2018-08-27 11:08	3.4 KB
unicom_cache_image	drwxrwx--x	2018-06-26 16:22	3.4 KB
UnicomLog	drwxrwx--x	2018-07-04 09:58	3.4 KB
userexperience	drwxrwx--x	2018-10-07 10:27	3.4 KB
websites	drwxrwx--x	2018-08-11 23:38	3.4 KB
WLAN Direct	drwxrwx--x	2018-05-27 23:34	3.4 KB
Xiaomi	drwxrwx--x	2018-09-14 13:00	3.4 KB
xtcdata	drwxrwx--x	2018-10-06 15:14	3.4 KB
zhibo8	drwxrwx--x	2018-06-09 10:15	3.4 KB
20181111112154.txt	-rw-rw----	2018-11-11 11:21	133 B
20181111150644.txt	-rw-rw----	2018-11-11 15:06	137 B
20181111150858.txt	-rw-rw----	2018-11-11 15:08	137 B
20181111151110.txt	-rw-rw----	2018-11-11 15:11	131 B
9c4955380c6ae21d	-rw-rw----	2018-08-10 20:20	22 B
com.qihoo360.mobilesafe.homepage	-rw-rw----	2018-10-30 15:13	0 B
config_system_switchs.txt	-rw-rw----	2018-06-29 09:12	5 B
d_1	-rw-rw----	2018-09-30 18:39	10.1 KB
FT_AkeySDK_2015.08.26.txt	-rw-rw----	2018-09-06 17:30	171 B
global_id.txt	-rw-rw----	2018-08-05 01:05	13 B
huawei.txt	-rw-rw----	2018-09-16 22:19	690 B
libCoreDaemon	-rw-rw----	2018-10-09 14:22	17.3 KB
movie.mp4	-rw-rw----	2018-09-20 18:26	8.3 MB
music.mp3	-rw-rw----	2018-09-20 18:26	10.3 MB
test.txt	-rw-rw----	2018-11-10 13:36	0 B
tmsdual_shark_mq.bat	-rw-rw----	2018-11-10 09:50	140 B
XTCWatch_5.2.1.apk	-rw-rw----	2018-08-27 11:10	49.9 MB
XTCWatch_5.3.6.apk	-rw-rw----	2018-09-07 16:11	52 MB
XTCWatch_5.4.7.apk	-rw-rw----	2018-11-01 17:05	51.4 MB

图 10-19　SD 卡查看文件

SharedPreference 文件读写（实践案例）

10.1.3　SharePreferences 存储

本小节我们介绍 SharePreferences 存储。SharePreferences 类是一个轻量级的存储类，特别适合保存软件配置参数。使用 SharedPreferences 保存数据，其背后是用 XML 文件存放数据，如图 10-20 所示。这种方式存储结构为 Key-Value 的键值对，主要用来存储比较简单的一些数据，而且数据类型是标准的 Boolean、Int、Float、Long、String 等类型。

```
<?xml version="1.0" encoding="UTF-8" standalone="true"?>
- <map>
    <string name="User">sziit</string>
    <string name="Password">123456</string>
  </map>
```

图 10-20　XML 文件

SharedPreferences 类存储文件的名字后无须考虑存储文件的后缀问题，因为 Android 系统会自动将你命名的文件后面加上.xml；在 Android 系统中，SharedPreferences 中的信息以 XML

文件的形式保存在 /data/data/PACKAGE_NAME/shared_prefs 目录下，如图 10-21 所示。

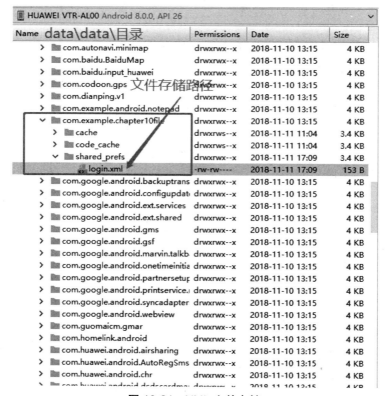

图 10-21　XML 文件存储

SharedPreference 文件有 4 种操作模式，4 种操作模式的说明如下。
◆ Context.MODE_PRIVATE：默认操作模式，代表该文件是私有数据，只能被应用本身访问。
◆ Context.MODE_APPEND：该模式会检查文件是否存在，存在就往文件中追加内容，否则就创建新文件。
◆ Context.MODE_WORLD_READABLE：表示当前文件可以被其他应用读取。
◆ Context.MODE_WORLD_WRITEABLE：表示当前文件可以被其他应用写入。

获取 SharedPreference 数据有以下两种方式。
◆ this.getPreferences (int mode)：调用 Activity 对象的 getPreferences()方法，通过 Activity 对象获取的是本 Activity 私有的 Preference。保存在系统中的 XML 形式的文件的名称为这个 Activity 的名字。
◆ this.getSharedPreferences (String name, int mode)：调用 Context 对象的 getPreferences()方法，调用 Context 对象的 getSharedPreferences()方法，以第一个参数的 name 为文件名保存在系统中。

下面我们看一个通过 SharedPreference 实现用户名和密码的存储的实例，如图 10-22 所示。用户输入用户名和密码后，选择"保存密码"复选框，单击"登录"按钮后，用户名和密码存储到 SharedPreference 文件中。

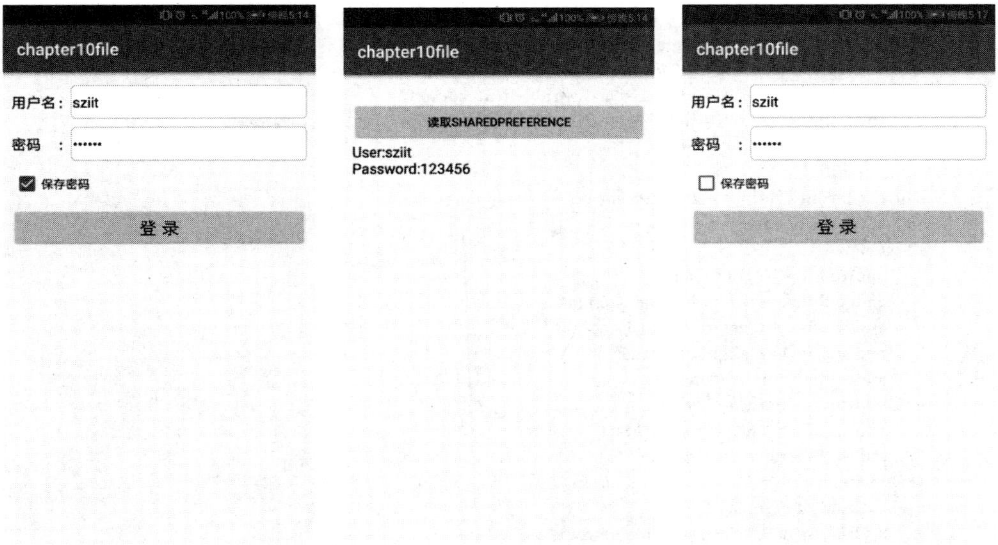

图 10-22 SharedPreference 存储

首先我们完成 SharedPreference 保存密码的界面布局 activity_sharedpreference_write.xml，如图 10-23 所示。

```xml
<LinearLayout xmlns:android="http://schemas.android.com/apk/res/android"
    android:layout_width="match_parent"
    android:layout_height="match_parent"
    android:focusable="true"
    android:focusableInTouchMode="true"
    android:orientation="vertical"
    android:padding="10dp" >
    <RelativeLayout
        android:layout_width="match_parent"
        android:layout_height="50dp" >
        <TextView
            android:id="@+id/tv_name"
            android:layout_width="wrap_content"
            android:layout_height="match_parent"
            android:layout_alignParentLeft="true"
            android:gravity="center"
            android:text="用户名："
            android:textColor="@color/black"
            android:textSize="17sp" />
        <EditText
            android:id="@+id/et_name"
            android:layout_width="match_parent"
            android:layout_height="match_parent"
            android:layout_marginLeft="0dp"
            android:layout_marginTop="5dp"
            android:layout_marginBottom="5dp"
            android:layout_toRightOf="@+id/tv_name"
            android:background="@drawable/edittext_selector"
            android:gravity="left|center"
            android:hint="请输入用户名"
            android:inputType="text"
            android:maxLength="12"
```

图 10-23 SharedPreference 保存界面布局

```xml
            android:textColor="@color/black"
            android:textColorHint="@color/grey"
            android:textCursorDrawable="@drawable/text_cursor"
            android:textSize="17sp" />
    </RelativeLayout>
    <RelativeLayout
        android:layout_width="match_parent"
        android:layout_height="50dp" >
        <TextView
            android:id="@+id/tv_no"
            android:layout_width="wrap_content"
            android:layout_height="match_parent"
            android:layout_alignParentLeft="true"
            android:gravity="center"
            android:text="密　码　：  "
            android:textColor="@color/black"
            android:textSize="17sp" />
        <EditText
            android:id="@+id/et_psd"
            android:layout_width="match_parent"
            android:layout_height="match_parent"
            android:layout_marginTop="5dp"
            android:layout_marginBottom="5dp"
            android:layout_toRightOf="@+id/tv_no"
            android:background="@drawable/editext_selector"
            android:gravity="left|center"
            android:hint="请输入密码"
            android:inputType="numberPassword"
            android:maxLength="20"
            android:textColor="@color/black"
            android:textColorHint="@color/grey"
            android:textCursorDrawable="@drawable/text_cursor"
            android:textSize="17sp" />
    </RelativeLayout>
    <RelativeLayout
        android:layout_width="match_parent"
        android:layout_height="50dp" >
        <CheckBox
            android:id="@+id/checkBox_Psd"
            android:layout_width="wrap_content"
            android:layout_height="wrap_content"
            android:layout_alignParentStart="true"
            android:layout_alignParentLeft="true"
            android:layout_alignParentBottom="true"
            android:layout_marginStart="2dp"
            android:layout_marginLeft="2dp"
            android:layout_marginBottom="12dp"
            android:text="保存密码" />
    </RelativeLayout>
    <Button
        android:id="@+id/btn_save"
        android:layout_width="match_parent"
        android:layout_height="wrap_content"
        android:text="登　录  "
        android:textColor="@color/black"
        android:textSize="20sp" />
</LinearLayout>
```

图 10-23　SharedPreference 保存界面布局（续）

SharedPreference 保存界面布局的效果如图 10-24 所示。

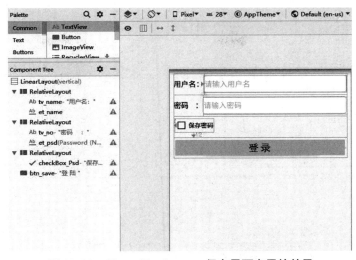

图 10-24　SharedPreference 保存界面布局的效果

下面我们看一下读取 SharedPreference 文件的实现过程。

① SharedPreference 文件的初始化，使用 getSharedPreferences(""login"", Context.MODE_PRIVATE)的方式读取文件，将内容存储到 login.xml 中，文件只能被应用本身访问，如图 10-25 所示。

```
private SharedPreferences mSharedPreferences;
private void initData() {
    mSharedPreferences = getSharedPreferences("login", Context.MODE_PRIVATE);
}
```

图 10-25　SharedPreference 文件初始化

② 键值对数据写入，如图 10-26 所示。mSharedPreferences.edit()获取文件的编辑权限，写入键值"User"和"Password"，将键值对写入文件。

```
if (mCheckBoxPsd.isChecked()) {
//获取文件的编辑权限
    SharedPreferences.Editor mEditor = mSharedPreferences.edit();        mEditor.putString("User", name);//写入键值"User"
    mEditor.putString("Password", psd);//写入键值"Password"
    mEditor.commit();//将键值对写入文件
}
```

图 10-26　键值对数据写入

新建 SharedPreference 保存视图 SharedWriteActivity，加载布局文件 activity_sharedpreference_write.xml。SharedPreference 保存视图 SharedWriteActivity 的整体代码如图 10-27 所示。

```
package com.example.chapter10file;
import android.content.Context;
import android.content.SharedPreferences;
import android.os.Bundle;
import android.support.annotation.Nullable;
```

图 10-27　SharedWriteActivity 代码实现

```java
import android.support.v7.app.AppCompatActivity;
import android.text.TextUtils;
import android.view.View;
import android.widget.Button;
import android.widget.CheckBox;
import android.widget.EditText;
import android.widget.Toast;
public class SharedWriteActivity extends AppCompatActivity implements View.OnClickListener {
    private EditText mEtName;
    private EditText mEtPsd;
    private CheckBox mCheckBoxPsd;
    private Button mBtnSave;
    private static SharedPreferences mSharedPreferences;
    @Override
    protected void onCreate(@Nullable Bundle savedInstanceState) {
        super.onCreate(savedInstanceState);
        setContentView(R.layout.activity_sharedpreference_write);
        initView();
        initData();
    }
    private void initData() {
//如果已保存数据的话,将数据读出并显示
        mSharedPreferences = getSharedPreferences("login", Context.MODE_PRIVATE);
        mEtName.setText(mSharedPreferences.getString("User", null));
        mEtPsd.setText(mSharedPreferences.getString("Password", null));
    }
    private void initView() {
        mEtName = (EditText) findViewById(R.id.et_name);
        mEtPsd = (EditText) findViewById(R.id.et_psd);
        mCheckBoxPsd = (CheckBox) findViewById(R.id.checkBox_Psd);
        mBtnSave = (Button) findViewById(R.id.btn_save);
        mBtnSave.setOnClickListener(this);
    }
    @Override
    public void onClick(View v) {
        switch (v.getId()) {
            case R.id.btn_save:
                submit();
                break;
        }
    }
    private void submit() {
        // validate
        String name = mEtName.getText().toString().trim();
        if (TextUtils.isEmpty(name)) {
            Toast.makeText(this, "请输入用户名", Toast.LENGTH_SHORT).show();
            return;
        }
        String psd = mEtPsd.getText().toString().trim();
        if (TextUtils.isEmpty(psd)) {
            Toast.makeText(this, "请输入密码", Toast.LENGTH_SHORT).show();
            return;
        }
        // TODO validate success, do something
        if (mCheckBoxPsd.isChecked()) {
            SharedPreferences.Editor mEditor = mSharedPreferences.edit();
            mEditor.putString("User", name);
            mEditor.putString("Password", psd);
            mEditor.commit();
            Toast.makeText(this, "密码已保存", Toast.LENGTH_SHORT).show();
        }
    }
}
```

图 10-27　SharedWriteActivity 代码实现(续)

下面我们完成 SharedPreference 文件读取的功能。读取的界面布局 activity_sharedpreference_read.xml 如图 10-28 所示。

```xml
<LinearLayout xmlns:android="http://schemas.android.com/apk/res/android"
    xmlns:tools="http://schemas.android.com/tools"
    android:layout_width="match_parent"
    android:layout_height="match_parent"
    android:focusable="true"
    android:focusableInTouchMode="true"
    android:orientation="vertical"
    android:padding="10dp" >
    <TextView
        android:id="@+id/tv_hint"
        android:layout_width="match_parent"
        android:layout_height="wrap_content"
        android:gravity="center"
        android:textColor="@color/black"
        android:textSize="17sp"
        tools:text="读取共享文件内容" />
    <Button
        android:id="@+id/button"
        android:layout_width="match_parent"
        android:layout_height="wrap_content"
        android:text="读取 SharedPreference" />
    <TextView
        android:id="@+id/tv_filecontent"
        android:layout_width="match_parent"
        android:layout_height="619dp"
        android:textColor="@color/black"
        android:textSize="17sp" />
</LinearLayout>
```

图 10-28　SharedPreference 读取的界面布局

SharedPreference 读取界面布局的效果如图 10-29 所示。

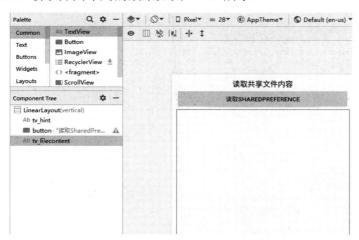

图 10-29　SharedPreference 读取界面布局的效果

读取键值对数据，SharedPreferences 进行初始化后，就可以读取键值"User"和"Password"，读取的过程如图 10-30 所示。

```java
private SharedPreferences mSharedPreferences;
private void initData() {
    mSharedPreferences = getSharedPreferences("login", Context.MODE_PRIVATE);
}
private void readData() {
```

图 10-30　SharedPreference 读取代码

```
        String strContent="";
        strContent+="User:"+mSharedPreferences.getString("User", null)+"\n";
        strContent+="Password:"+mSharedPreferences.getString("Password", null)+"\n";
        mTvFilecontent.setText(strContent);
    }
```

图 10-30　SharedPreference 读取代码（续）

新建 SharedPreference 读取视图 SharedReadActivity，加载布局文件 activity_sharedpreference_read.xml。SharedPreference 读取视图 SharedReadActivity 的整体代码如图 10-31 所示。

```
package com.example.chapter10file;
import android.content.Context;
import android.content.SharedPreferences;
import android.os.Bundle;
import android.support.annotation.Nullable;
import android.support.v7.app.AppCompatActivity;
import android.view.View;
import android.widget.Button;
import android.widget.TextView;
public class SharedReadActivity extends AppCompatActivity implements View.OnClickListener {
    private Button mButton;
    private TextView mTvFilecontent;
    private SharedPreferences mSharedPreferences;
    @Override
    protected void onCreate(@Nullable Bundle savedInstanceState) {
        super.onCreate(savedInstanceState);
        setContentView(R.layout.activity_sharedpreference_read);
        initView();
        initData();
    }
    private void initData() {
        mSharedPreferences = getSharedPreferences("login", Context.MODE_PRIVATE);
    }
    private void initView() {
        mButton = (Button) findViewById(R.id.button);
        mTvFilecontent = (TextView) findViewById(R.id.tv_filecontent);
        mButton.setOnClickListener(this);
    }
    @Override
    public void onClick(View v) {
        switch (v.getId()) {
            case R.id.button:
                readData();
                break;
        }
    }
    private void readData() {
        String strContent="";
        strContent+="User:"+mSharedPreferences.getString("User", null)+"\n";
        strContent+="Password:"+mSharedPreferences.getString("Password", null)+"\n";
        mTvFilecontent.setText(strContent);
    }
}
```

图 10-31　SharedReadActivity 读取代码

10.1.4　单元小测

判断题：

1. SharedPreference 将数据存储到 XML 文件中，以 Map<key,value>形式保存。（　　）

A. 是 　　　　　　　　　　　　　B. 否

2. 文件存储通过 I/O 数据流的方式将数据存储到文件中。（　　）

A. 是 　　　　　　　　　　　　　B. 否

3 Content Provider 存储主要用于显示程序中的数据。（　　）

A. 是 　　　　　　　　　　　　　B. 否

4. 文件存储将数据保存到 SD 卡时，用户需要在清单文件中增加权限"android.permission.READ_EXTERNAL_STORAGE"。（　　）

A 是 　　　　　　　　　　　　　B 否

选择题：

1. 下列文件的操作权限中，指定文件内容可以追加的是（　　）。

A. Context.MODE_PRIVATE

B. Context.MODE_APPEND

C. Context.MODE_WORLD_READABLE

D. Context.MODE_WORLD_WRITEABLE

2. 下列文件的操作权限中，指定文件内容只能被应用本身访问的是（　　）。

A. Context.MODE_PRIVATE

B. Context.MODE_APPEND

C. Context.MODE_WORLD_READABLE

D. Context.MODE_WORLD_WRITEABLE

3. 下列文件的操作权限中，指定文件内容可以被其他应用写入的是（　　）。

A. Context.MODE_PRIVATE

B. Context.MODE_APPEND

C. Context.MODE_WORLD_READABLE

D. Context.MODE_WORLD_WRITEABLE

4. 下列文件的操作权限中，指定文件内容可以被其他应用读取的是（　　）。

A. Context.MODE_PRIVATE

B. Context.MODE_APPEND

C. Context.MODE_WORLD_READABLE

D. Context.MODE_WORLD_WRITEABLE

5. Enviroment 类用于获取 SD 卡路径的是（　　）。

A. getSDStorageDirectory()

B. getSDStorageState()

C. getExternalStorageDirectory()

D. getExternalStorageState()

6. Enviroment 类用于获取 SD 卡状态的是（　　）。

A. getSDStorageDirectory()

B. getSDStorageState()

C. getExternalStorageDirectory()

D. getExternalStorageState()

7. 下列关于存储数据的说法中错误的是（　　）。
A. 文件存储以流的方式操作数据
B. 文件存储可以将数据存储到内存
C. 文件存储可以将数据存储到 SD 卡
D. Android 只能使用文件存储

8. 将程序中的私有数据分享给其他的应用程序，可以使用的是（　　）。
A. 文件存储　　　　　　　　　　　　B. 数据库存储
C. Content Provider 存储　　　　　　D. Sqlite 轻量级存储

9. 下面代码用于实现文件的保存，请补全代码。（　　）

```
{
FileOutputStream fos = new FileOutputStream(path);
(　?　);
fos.close();
}
```

A. fos.writeToFile(txt.getBytes());
B. fos.write(txt.getBytes());
C. fos.saveToFile(txt.getBytes());
D. fos.save(txt.getBytes());

10. 下面代码用于实现文件的读取，请补全代码。（　　）

```
{
{FileInputStream fis = new FileInputStream(path);
byte[] b = new byte[fis.available()];
(　?　)
readStr = new String(b);fis.close();    }
```

A. fis.input(b)　　　　　　　　　　　B. fis.readFromFile(b)
C. fis.read(b)　　　　　　　　　　　D. fis.getFromFile(b)

11. 下面代码用于实现 XML 文件的写入，请补全代码。（　　）

```
{
SharedPreferences.Editor mEditor = mSharedPreferences.edit();//获取文件的编辑权限        mEditor.putString("User", name);//写入键值"User"
mEditor.putString("Password", psd);//写入键值"Password"
(　?　);
}
```

A. mEditor.add()　　B. mEditor.execute()　　C. mEditor.enqueue()　　D. mEditor.commit()

10.2　数据库存储

10.2.1　知识点讲解——嵌入式数据库 SQLite

数据库存储（慕课）

上一节我们学习了 Android 的文件存储和 SharedPreference 存储，这些数据存储方式可以满足我们日常开发中存储少量数据的需求。如果使用它们存储一些数据量较大并且逻辑关系较为复杂的数据集，它们便显得较为笨拙和效率低下。那有没有更好的存储方案来解决此类问题呢？Google 为 Android 系统内置轻便又功能强大的嵌入式数据库 SQLite。本小节我们将会深入地学习如何在 Android 中使用 SQLite 数据库存储数据。在学习使用 SQLite 之前，我们先来简单了解一下 SQLite 数据库的特点。

◆ 轻量级：只需要一个动态库就可以享受它的全部功能，动态库的尺寸比较小。
◆ 独立性：SQLite 数据库的核心引擎不需要依赖第三方软件。
◆ 隔离性：SQLite 数据库中所有的信息（比如表、视图、触发器等）都包含在一个文件夹内，方便管理和维护。
◆ 跨平台：SQLite 目前支持大部分操作系统，可以在计算机操作系统和多个手机系统中运行，比如，Android 和 iOS。
◆ 安全性：SQLite 数据库通过数据库级上的独占性和共享锁来实现独立事务处理。这意味着多个进程可以在同一时间从同一数据库读取数据，但只能有一个可以写入数据。
◆ 多语言接口：SQLite 数据库支持多语言编程接口。

10.2.2 实践案例——将个人信息存储到 SQLite 数据库

简单地了解了 SQLite 数据库后，我们再来学习一下如何在 Android 中编写应用程序来使用 SQLite 执行数据的操作。我们通过一个例子来完成学习任务：实现的功能就是将一些个人信息存储到 SQLite 数据库，根据姓名删除数据库中的记录；根据姓名在数据库中查询记录并显示，再根据姓名查询数据库中的记录并更改记录，如图 10-32 所示。

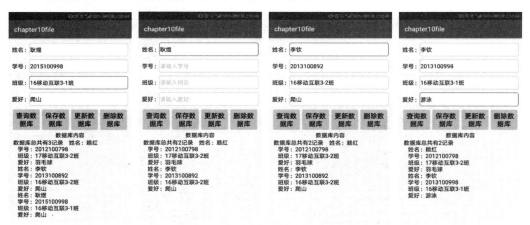

图 10-32　SQLite 数据库操作实例

Android 提供了 SQLiteOpenHelper 方法进行数据库的创建和升级，用户基于 SQLiteOpenHelper 来创建数据库。SQLiteOpenHelper 类方法如表 10-2 所示。

表 10-2　SQLiteOpenHelper 类的方法

序号	类方法	方法说明
1	MySQLiteOpenHelper 构造函数	新建一个数据库
2	onCreate(SQLiteDatabase db)	新建一个数据库表
3	onUpgrade(SQLiteDatabase db, int oldVersion, int newVersion)	数据库的升级使用

SQLite 数据库操作布局（实践案例）

首先我们完成数据库存储实例的界面布局 activity_sqlite.xml，如图 10-33 所示。

```xml
<LinearLayout xmlns:android="http://schemas.android.com/apk/res/android"
    android:layout_width="match_parent"
    android:layout_height="match_parent"
    android:focusable="true"
    android:focusableInTouchMode="true"
    android:orientation="vertical"
    android:padding="10dp" >
    <RelativeLayout
        android:layout_width="match_parent"
        android:layout_height="50dp" >
        <TextView
            android:id="@+id/tv_name"
            android:layout_width="wrap_content"
            android:layout_height="match_parent"
            android:layout_alignParentLeft="true"
            android:gravity="center"
            android:text="姓名:"
            android:textColor="@color/black"
            android:textSize="17sp" />
        <EditText
            android:id="@+id/et_name"
            android:layout_width="match_parent"
            android:layout_height="match_parent"
            android:layout_marginBottom="5dp"
            android:layout_marginTop="5dp"
            android:layout_toRightOf="@+id/tv_name"
            android:background="@drawable/editext_selector"
            android:gravity="left|center"
            android:hint="请输入姓名"
            android:inputType="text"
            android:maxLength="12"
            android:textColor="@color/black"
            android:textColorHint="@color/grey"
            android:textCursorDrawable="@drawable/text_cursor"
            android:textSize="17sp" />
    </RelativeLayout>
    <RelativeLayout
        android:layout_width="match_parent"
        android:layout_height="50dp" >
        <TextView
            android:id="@+id/tv_no"
            android:layout_width="wrap_content"
            android:layout_height="match_parent"
            android:layout_alignParentLeft="true"
            android:gravity="center"
            android:text="学号:"
            android:textColor="@color/black"
            android:textSize="17sp" />
        <EditText
            android:id="@+id/et_no"
            android:layout_width="match_parent"
            android:layout_height="match_parent"
            android:layout_marginTop="5dp"
            android:layout_marginBottom="5dp"
            android:layout_toRightOf="@+id/tv_no"
            android:background="@drawable/editext_selector"
            android:gravity="left|center"
            android:hint="请输入学号"
            android:inputType="text"
            android:maxLength="20"
            android:textColor="@color/black"
            android:textColorHint="@color/grey"
            android:textCursorDrawable="@drawable/text_cursor"
            android:textCursorDrawable="@drawable/text_cursor"
            android:textSize="17sp" />
    </RelativeLayout>
```

图 10-33 数据库实例界面布局

```xml
<RelativeLayout
    android:layout_width="match_parent"
    android:layout_height="50dp" >
    <TextView
        android:id="@+id/tv_class"
        android:layout_width="wrap_content"
        android:layout_height="match_parent"
        android:layout_alignParentLeft="true"
        android:gravity="center"
        android:text="班级："
        android:textColor="@color/black"
        android:textSize="17sp" />
    <EditText
        android:id="@+id/et_class"
        android:layout_width="match_parent"
        android:layout_height="match_parent"
        android:layout_marginTop="5dp"
        android:layout_marginBottom="5dp"
        android:layout_toRightOf="@+id/tv_class"
        android:background="@drawable/edittext_selector"
        android:gravity="left|center"
        android:hint="请输入班级"
        android:inputType="text"
        android:maxLength="20"
        android:textColor="@color/black"
        android:textColorHint="@color/grey"
        android:textCursorDrawable="@drawable/text_cursor"
        android:textSize="17sp" />
</RelativeLayout>
<RelativeLayout
    android:layout_width="match_parent"
    android:layout_height="50dp" >
    <TextView
        android:id="@+id/tv_hobby"
        android:layout_width="wrap_content"
        android:layout_height="match_parent"
        android:layout_alignParentLeft="true"
        android:gravity="center"
        android:text="爱好："
        android:textColor="@color/black"
        android:textSize="17sp" />
    <EditText
        android:id="@+id/et_hobby"
        android:layout_width="match_parent"
        android:layout_height="match_parent"
        android:layout_marginTop="5dp"
        android:layout_marginBottom="5dp"
        android:layout_toRightOf="@+id/tv_hobby"
        android:background="@drawable/edittext_selector"
        android:gravity="left|center"
        android:hint="请输入爱好:"
        android:inputType="text"
        android:maxLength="20"
        android:textColor="@color/black"
        android:textColorHint="@color/grey"
        android:textCursorDrawable="@drawable/text_cursor"
        android:textSize="17sp" />
</RelativeLayout>
<LinearLayout
    android:layout_width="match_parent"
    android:layout_height="wrap_content"
    android:orientation="horizontal">
    <Button
        android:id="@+id/btn_query"
        android:layout_width="0dp"
        android:layout_height="wrap_content"
        android:layout_weight="1"
        android:text="查询数据库"
```

图 10-33　数据库实例界面布局（续）

```xml
            android:textColor="@color/black"
            android:textSize="20sp" />
        <Button
            android:id="@+id/btn_save"
            android:layout_width="0dp"
            android:layout_height="wrap_content"
            android:layout_weight="1"
            android:text="保存数据库"
            android:textColor="@color/black"
            android:textSize="20sp" />

        <Button
            android:id="@+id/btn_update"
            android:layout_width="0dp"
            android:layout_height="wrap_content"
            android:layout_weight="1"
            android:text="更新数据库"
            android:textColor="@color/black"
            android:textSize="20sp" />
        <Button
            android:id="@+id/btn_delete"
            android:layout_width="0dp"
            android:layout_height="wrap_content"
            android:layout_weight="1"
            android:text="删除数据库"
            android:textColor="@color/black"
            android:textSize="20sp" />
    </LinearLayout>
    <TextView
        android:layout_width="match_parent"
        android:layout_height="wrap_content"
        android:gravity="center"
        android:text="数据库内容"
        android:textColor="@color/black"
        android:textSize="17sp" />
    <TextView
        android:id="@+id/tv_database"
        android:layout_width="match_parent"
        android:layout_height="match_parent"
        android:layout_gravity="center"
        android:textColor="@color/black"
        android:textSize="17sp" />
</LinearLayout>
```

图 10-33　数据库实例界面布局（续）

数据库实例界面布局的效果如图 10-34 所示。

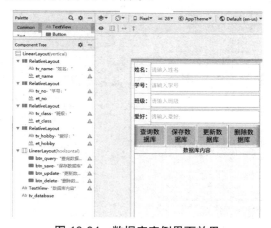

图 10-34　数据库实例界面效果

我们看一下如何在数据库中创建一个表,比如我们创建一个数据库 Student.db,在数据库 Student.db 表下新建一个数据表 user。在 Android 中我们使用 SQL 变量 CREATE_USER 来创建表,使用 create table user 这条 SQL 语句来创建数据库表,如图 10-35 所示。

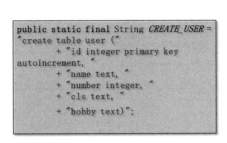
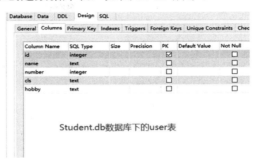

图 10-35　SQLite 数据库创建

SQLiteOpenHelper 实现过程如下,实现代码如图 10-36 所示。

① 用户基于 SQLiteOpenHelper 创建子类 MySQLiteOpenHelper,实现构造函数 MySQLite OpenHelper(),将视图全局变量赋值给 mCtx。

② 通过 SQL 语句创建数据库表,使用 db.execSQL(CREATE_USER) 创建数据库表。

③ 更改版本号升级数据,使用 db.execSQL(""drop table if exists user"")删除旧表,使用 onCreate(db)创建新表。

SQLite 数据库操作
(实践案例)

```
package com.example.chapter10file.db;
import android.content.Context;
import android.database.sqlite.SQLiteDatabase;
import android.database.sqlite.SQLiteOpenHelper;
import android.widget.Toast;
public class MySQLiteOpenHelper extends SQLiteOpenHelper {
    public static final String CREATE_USER = "create table user ("
            + "id integer primary key autoincrement, "
            + "name text, "
            + "number integer, "
            + "cls text, "
            + "hobby text)";
    private Context mCtx;
    public MySQLiteOpenHelper(Context context, String name, SQLiteDatabase.CursorFactory factory, int version) {
        super(context, name, factory, version);
        this.mCtx=context;
    }
    @Override
    public void onCreate(SQLiteDatabase db) {
        db.execSQL(CREATE_USER);
        Toast.makeText(mCtx, "数据库表 user 创建成功", Toast.LENGTH_SHORT).show();
    }
    @Override
    public void onUpgrade(SQLiteDatabase db, int oldVersion, int newVersion) {
        db.execSQL("drop table if exists user");
        onCreate(db);
    }
}
```

图 10-36　MySQLiteOpenHelper 类

SQLiteOpenHelper 使用 getWritableDatabase 方法获取 SQLiteDatabase 接口,mSQLiteDatabase= mDbHelper.getWritableDatabase()。利用 mSQLiteDatabase 就可以进行数据库的操作。

SQLiteDatabase 操作数据库的接口如表 10-3 所示。

表 10-3　SQLiteDatabase 操作数据库的接口

序号	类方法	方法说明
1	insert(String table, String nullColumnHack, ContentValues values)	向数据库中插入一条数据
2	delete(String table, String whereClause, String[] whereArgs)	根据查询条件删除数据
3	update(String table, ContentValues values, String whereClause, String[] whereArgs)	根据查询条件更新数据
4	query(String table, String[] columns, String selection, String[] selectionArgs, String groupBy, String having, String orderBy)	根据条件查询数据

数据库 Cursor 使用方法如表 10-4 所示。

表 10-4　Cursor 使用方法

序号	类方法	方法说明
1	moveToFirst	移动游标到开头
2	moveToNext	移动游标到下一条记录
3	moveToLast	移动游标到结尾
4	getCount	获取查询结果的总数量
5	getInt getString	获取指定字段的值

对数据库的操作需要建立一个数据 Bean 类 UserInfo，如图 10-37 所示，主要包括"姓名""学号""班级""爱好"变量，以及变量属性的 GET 和 SET 方法。

```java
package com.example.chapter10file.bean;
public class UserInfo {
    private String sName;
    private int iNumber;
    private String strClass;
    private String strHobby;
    public String getsName() {
        return sName;
    }
    public void setsName(String sName) {
        this.sName = sName;
    }
    public int getiNumber() {
        return iNumber;
    }
    public void setiNumber(int iNumber) {
        this.iNumber = iNumber;
    }
    public String getStrClass() {
        return strClass;
    }
    public void setStrClass(String strClass) {
        this.strClass = strClass;
    }
    public String getStrHobby() {
        return strHobby;
    }
    public void setStrHobby(String strHobby) {
        this.strHobby = strHobby;
    }
}
```

图 10-37　UserInfo 类

增加数据库记录的流程如图 10-38 所示，实现的流程如下：
- 新建数据对象。
- 分别存储数据库中的"姓名""学号""班级""爱好"字段。
- 将数据存储到数据库中。

```java
private void insert() {
    ContentValues mContentValues = new ContentValues();//新建数据对象
    mContentValues.put("name", mUserInfo.getsName());//存储"姓名"字段
    mContentValues.put("number", mUserInfo.getiNumber()); //存储"学号"字段
    mContentValues.put("cls", mUserInfo.getStrClass()); //存储"班级"字段
    mContentValues.put("hobby", mUserInfo.getStrHobby()); //存储"爱好"字段
    SQLiteDatabase.insert("user", null, mContentValues); //将数据存储到数据库
}
```

图 10-38　增加数据库记录

删除数据库记录的流程如图 10-39 所示。
- 首先获取查询姓名。
- 根据姓名查询数据库中的记录并删除记录。

```java
private void delete() {
    String name = mEtName.getText().toString().trim();//获取查询姓名
    if (TextUtils.isEmpty(name)) {
        Toast.makeText(this, "请输入姓名", Toast.LENGTH_SHORT).show();
        return;
    }
    //根据姓名查询数据库并删除记录
    mSQLiteDatabase.delete("user", "name=?", new String[]{name});
}
```

图 10-39　删除数据库记录

更新数据库记录的流程如图 10-40 所示。
- 新建数据对象。
- 分别更新数据库中的"姓名""学号""班级""爱好"字段。
- 根据姓名查询数据库中的记录并更新数据库记录。

```java
private void insert() {
    ContentValues mContentValues = new ContentValues(); //新建数据对象
    mContentValues.put("name", mUserInfo.getsName()); //更新"姓名"字段
    mContentValues.put("number", mUserInfo.getiNumber()); //更新"学号"字段
    mContentValues.put("cls", mUserInfo.getStrClass()); //更新"班级"字段
    mContentValues.put("hobby", mUserInfo.getStrHobby()); //更新"爱好"字段
    //根据姓名查询数据库并更新数据库记录
    mSQLiteDatabase.update("user", mContentValues,"name=?",newString[]{mUserInfo.getsName()});
}
```

图 10-40　更新数据库记录

增加记录后的数据库如图 10-41 所示。

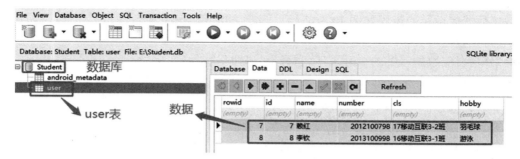

图 10-41 数据库记录表

查询数据库记录的流程如图 10-42 所示。
◆ 根据姓名查询数据库。
◆ 如果没有姓名,则查询所有数据库。
◆ 根据数据库游标轮询所有数据,取出记录中的每一条数据。

```java
private void query() {
    String name = mEtName.getText().toString().trim();
    //1 根据姓名查询
    Cursor cursor = mSQLiteDatabase.query("user", null,"name=?", new String[]{name},null,null,null);
    //2 查询所有数据
    Cursor cursor = mSQLiteDatabase.query("user", null, null, null,null,null);
    if(cursor.moveToFirst())
    {
        do {
            //3 轮询所有数据,取出每一条数据
            String strHobby = cursor.getString(cursor.getColumnIndex("hobby"));
            mEtHobby.setText(strHobby);
            return ;
        }while (cursor.moveToNext());
    }
    return ;
}
```

图 10-42 查询数据库记录

新建数据库操作视图 SqliteActivity,加载布局文件 activity_sqlite.xml。数据库操作视图 SqliteActivity 的整体代码如图 10-43 所示。

```java
package com.example.chapter10file;
import android.content.ContentValues;
import android.database.Cursor;
import android.database.sqlite.SQLiteDatabase;
import android.os.Bundle;
import android.support.v7.app.AppCompatActivity;
import android.text.TextUtils;
import android.view.View;
import android.widget.Button;
import android.widget.EditText;
import android.widget.TextView;
import android.widget.Toast;
import com.example.chapter10file.bean.UserInfo;
import com.example.chapter10file.db.MySQLiteOpenHelper;
public class SqliteActivity extends AppCompatActivity implements View.OnClickListener {
    private EditText mEtName;
    private EditText mEtNo;
    private EditText mEtClass;
```

图 10-43 数据库操作视图 SqliteActivity

```java
    private EditText mEtHobby;
    private Button mBtnQuery;
    private Button mBtnSave;
    private Button mBtnUpdate;
    private Button mBtnDelete;
    private TextView mTvDatabase;
    private MySQLiteOpenHelper mDbHelper;
    private SQLiteDatabase mSQLiteDatabase;
    private final String DB_NAME = "Student.db";
    UserInfo mUserInfo =new UserInfo();
    @Override
    protected void onCreate(Bundle savedInstanceState) {
        super.onCreate(savedInstanceState);
        setContentView(R.layout.activity_sqlite);
        initView();
        initData();
    }
    private void initData() {
        mDbHelper = new MySQLiteOpenHelper(this, DB_NAME, null, 1);
        mSQLiteDatabase = mDbHelper.getWritableDatabase();
        queryAll();
    }
    private void initView() {
        mEtName = (EditText) findViewById(R.id.et_name);
        mEtNo = (EditText) findViewById(R.id.et_no);
        mEtClass = (EditText) findViewById(R.id.et_class);
        mEtHobby = (EditText) findViewById(R.id.et_hobby);
        mBtnQuery = (Button) findViewById(R.id.btn_query);
        mBtnSave = (Button) findViewById(R.id.btn_save);
        mBtnUpdate = (Button) findViewById(R.id.btn_update);
        mBtnDelete = (Button) findViewById(R.id.btn_delete);
        mTvDatabase = (TextView) findViewById(R.id.tv_database);
        mBtnQuery.setOnClickListener(this);
        mBtnSave.setOnClickListener(this);
        mBtnUpdate.setOnClickListener(this);
        mBtnDelete.setOnClickListener(this);
    }
    @Override
    public void onClick(View v) {
        switch (v.getId()) {
            case R.id.btn_query:
                query();
                break;
            case R.id.btn_save:
                insert();
                break;
            case R.id.btn_update:
                update();
                break;
            case R.id.btn_delete:
                delete();
                break;
        }
    }
    /*向数据库增加一条记录*/
    private void insert() {
        submit();
        ContentValues mContentValues = new ContentValues();
        mContentValues.put("name", mUserInfo.getsName());
        mContentValues.put("number", mUserInfo.getiNumber());
        mContentValues.put("cls", mUserInfo.getStrClass());
        mContentValues.put("hobby", mUserInfo.getStrHobby());
        mSQLiteDatabase.insert("user", null, mContentValues);
        Toast.makeText(this, " 插入数据成功", Toast.LENGTH_SHORT).show();
        queryAll();
```

图 10-43　数据库操作视图 SqliteActivity（续）

```java
}
/*数据库删除一条记录*/
private void delete() {
    String name = mEtName.getText().toString().trim();
    if (TextUtils.isEmpty(name)) {
        Toast.makeText(this, "请输入姓名", Toast.LENGTH_SHORT).show();
        return;
    }
    mSQLiteDatabase.delete("user", "name=?", new String[]{name});
    Toast.makeText(this, "  删除成功", Toast.LENGTH_SHORT).show();
    queryAll();
}

/*数据库更新一条记录*/
private void update() {
    submit();
    ContentValues mContentValues = new ContentValues();
    mContentValues.put("name", mUserInfo.getsName());
    mContentValues.put("number", mUserInfo.getiNumber());
    mContentValues.put("cls", mUserInfo.getStrClass());
    mContentValues.put("hobby", mUserInfo.getStrHobby());
    mSQLiteDatabase.update("user", mContentValues,"name=?",new String[]{mUserInfo.getsName()});
    Toast.makeText(this, "更新数据成功", Toast.LENGTH_SHORT).show();
    queryAll();
}
/*数据库查询一条记录*/
private void query() {
    String name = mEtName.getText().toString().trim();
    if (TextUtils.isEmpty(name)) {
        Toast.makeText(this, "请输入姓名", Toast.LENGTH_SHORT).show();
        return;
    }
    Cursor cursor = mSQLiteDatabase.query("user", null,"name=?", new String[]{name},null,null,null);
    if(cursor.moveToFirst())
    {
        do {
            String strName = cursor.getString(cursor.getColumnIndex("name"));
            int iNumber = cursor.getInt(cursor.getColumnIndex("number"));
            String strCls = cursor.getString(cursor.getColumnIndex("cls"));
            String strHobby = cursor.getString(cursor.getColumnIndex("hobby"));
            mEtNo.setText(String.valueOf(iNumber));
            mEtClass.setText(strCls);
            mEtHobby.setText(strHobby);
            Toast.makeText(this, "  查询成功", Toast.LENGTH_SHORT).show();
            return ;
        }while (cursor.moveToNext());
    }
    Toast.makeText(this, "  查询失败", Toast.LENGTH_SHORT).show();
    return ;
}
/*检查 EditText 是否空白,并将 EditText 内容全部存入数据类对象*/
private void submit() {
    // validate
    String name = mEtName.getText().toString().trim();
    if (TextUtils.isEmpty(name)) {
        Toast.makeText(this, "请输入姓名", Toast.LENGTH_SHORT).show();
        return;
    }
    String no = mEtNo.getText().toString().trim();
    if (TextUtils.isEmpty(no)) {
        Toast.makeText(this, "请输入学号", Toast.LENGTH_SHORT).show();
        return;
    }
    String cls = mEtClass.getText().toString().trim();
```

图 10-43　数据库操作视图 SqliteActivity（续）

```
        if (TextUtils.isEmpty(cls)) {
            Toast.makeText(this, "请输入班级", Toast.LENGTH_SHORT).show();
            return;
        }
        String hobby = mEtHobby.getText().toString().trim();
        if (TextUtils.isEmpty(hobby)) {
            Toast.makeText(this, "请输入爱好:", Toast.LENGTH_SHORT).show();
            return;
        }
        mUserInfo.setsName(name);
        mUserInfo.setStrHobby(hobby);
        mUserInfo.setStrClass(cls);
        mUserInfo.setiNumber(Integer.parseInt(no));
    }
    /*查询数据库所有的记录*/
    public void queryAll() {
        String content = "";
        Cursor cursor = mSQLiteDatabase.query("user", null, null, null, null, null, null);
        cursor.getCount();
        content+=" 数据库总共有"+cursor.getCount()+"记录\n";
        while (cursor.moveToNext()) {
            String strName = cursor.getString(cursor.getColumnIndex("name"));
            int iNumber = cursor.getInt(cursor.getColumnIndex("number"));
            String strCls = cursor.getString(cursor.getColumnIndex("cls"));
            String strHobby = cursor.getString(cursor.getColumnIndex("hobby"));
            content = String.format("%s 姓名: %s\n", content, strName);
            content = String.format("%s 学号: %s\n", content, iNumber);
            content = String.format("%s 班级: %s\n", content, strCls);
            content = String.format("%s 爱好: %s\n", content, strHobby);
        }
        mTvDatabase.setText(content);
    }
}
```

图 10-43　数据库操作视图 SqliteActivity（续）

10.2.3　单元小测

判断题：

1. SQLite 数据库使用后不需要关闭，不影响性能。（　　）
   A. 是　　　　　　　　　　　　　　B. 否
2. SQLite 支持 SQL 语句的增、删、改、查等操作。（　　）
   A. 是　　　　　　　　　　　　　　B. 否

选择题：

1. SQLite 具备下面哪些特点？（　　）
   A. 轻量级　　　　B. 独立性　　　　C. 跨平台　　　　D. 多语言接口
2. SQLiteOpenHelper 类通过（　　）方法可以创建一个可写的数据库对象。
   A. getReadableDatabase()　　　　　　B. getWritableDatabase()
   C. getDatabase()　　　　　　　　　　D. getSqliteDatabase()
3. SQLiteOpenHelper 类通过（　　）方法可以创建一个可读的数据库对象。
   A. getDatabase()　　　　　　　　　　B. getSqliteDatabase()

C. getReadableDatabase() D. getSqliteDatabase()

4. Cursor 游标的（　　）方法可以移动游标到数据库的开头。

A. moveFirst()　　B. moveToFirst()　　C. moveStart()　　D. moveToStart()

5. Cursor 游标的（　　）方法可以移动游标到数据库的结尾。

A. moveEnd()　　B. moveToEnd()　　C. moveLast()　　D. moveToLast()

6. 下面代码用于向数据库增加一条记录，请补全代码。（　　）

```
private void insert() {
ContentValues mContentValues = new ContentValues();
mContentValues.put("name", mUserInfo.getsName());
mContentValues.put("number", mUserInfo.getiNumber());
mContentValues.put("cls", mUserInfo.getStrClass());
mContentValues.put("hobby", mUserInfo.getStrHobby());
(    ?    ) ;
}
```

A. SQLiteDatabase.add("user", null, mUserInfo)

B. SQLiteDatabase.add("user", null, mContentValues)

C. SQLiteDatabase.insert("user", null, mUserInfo)

D. SQLiteDatabase.insert("user", null, mContentValues)

7. 下面代码用于将数据库删除一条记录，请补全代码。（　　）

```
private void delete() {
String name = mEtName.getText().toString().trim();
(    ?    ) ;
}
```

A. mSQLiteDatabase.del("user", "name=?", new String[]{name})

B. mSQLiteDatabase.del("user", "name=?", new String{name})

C. mSQLiteDatabase.delete("user", "name=?", new String[]{name})

D. mSQLiteDatabase.delete("user", "name=?", new String{name})

8. 下面代码用于在数据库中更新一条记录，请补全代码。（　　）

```
private void insert() {
ContentValues mContentValues = new ContentValues();
mContentValues.put("name", mUserInfo.getsName());
mContentValues.put("number", mUserInfo.getiNumber());
mContentValues.put("cls", mUserInfo.getStrClass());
mContentValues.put("hobby", mUserInfo.getStrHobby());
(    ?    ) ;
}
```

A. mSQLiteDatabase.update("user", mContentValues,"name=?",newString{mUserInfo.getsName()})

B. mSQLiteDatabase.update("user", mContentValues,"name=?",newString[]{mUserInfo.getsName()})

C. mSQLiteDatabase.refresh("user", mContentValues,"name=?",newString{mUserInfo.getsName()})

D. mSQLiteDatabase.refresh("user", mContentValues,"name=?",newString[]{mUserInfo.getsName()})

本章课后练习和程序源代码

第 10 章源代码及课后习题

参考文献

[1] 郭霖. 第一行代码 Android[M]. 2 版. 北京：人民邮电出版社，2016.

[2] 传智播客. Android 移动应用基础教程（Android Studio）[M]. 2 版. 北京：中国铁道出版社，2019.

[3] 唐亮杜，杜秋阳. 用微课学 Android 开发基础[M]. 北京：高等教育出版社，2016.

[4] 唐亮杜，靳幸福. 用微课学 Android 高级开发[M]. 北京：高等教育出版社，2016.

[5] 欧阳燚. Android Studio 开发实战：从零基础到 App 上线[M]. 2 版. 北京：清华大学出版社，2018.

[6] 毕小朋. 精通 Android Studio[M]. 北京：清华大学出版社，2016.

[7] 肖琨，等. Android Studio 移动开发教程[M]. 北京：电子工业出版社，2019.

[8] 刘望舒. Android 进阶之光[M]. 北京：电子工业出版社，2017.

[9] 施威铭. Android App 开发入门：使用 Android Studio 2.X 开发环境[M]. 2 版. 北京：机械工业出版社，2017.

[10] 范磊. Android 应用开发进阶[M]. 北京：电子工业出版社，2018.

[11] 陈承欢. Android 移动应用开发任务驱动教程[M]. 北京：电子工业出版社，2016.

[12] 李瑞奇. 移动开发丛书 Android 开发实战：从学习到产品（适用于 Android 6/7 与 Android Studio 2.x）[M]. 北京：清华大学出版社，2017.

[13] 张光河. Android 移动开发案例教程——基于 Android Studio 开发环境[M]. 北京：清华大学出版社，2017.

[14] 郑丹青. Android 模块化开发项目式教程（Android Studio）[M]. 北京：人民邮电出版社，2018.

[15] 张思民. Android Studio 应用程序设计（微课版）[M]. 2 版. 北京：清华大学出版社，2017.10；

[16] 明日学院. Android 开发从入门到精通—项目案例版 Android Studio 软件编程应用设计[M]. 北京：中国水利水电出版社，2017.

[17] 方欣. Android Studio 应用开发—基础入门与应用实战[M]. 北京：电子工业出版社，2017.

[18] http://www.android-studio.org/，Android Studio 安卓开发者社区.

[19] https://www.csdn.net/；CSDN-专业 IT 技术社区.

[20] https://developer.android.google.cn/；Google Android Studio 开发者官方社区.